AFTERNOTES on NUMERICAL ANALYSIS

G. W. Stewart
University of Maryland
College Park, Maryland

AFTERNOTES
on
NUMERICAL
ANALYSIS

A series of lectures on elementary numerical analysis presented at the University of Maryland at College Park and recorded after the fact.

siam.
Society for Industrial and Applied Mathematics

Philadelphia

QA
297
S785
1996

Copyright © 1996 by the Society for Industrial and Applied Mathematics.

10 9 8 7 6 5 4 3 2 1

All rights reserved. Printed in the United States of America. No part of this book may be reproduced, stored, or transmitted in any manner without the written permission of the publisher. For information, write to the Society for Industrial and Applied Mathematics, 3600 University City Science Center, Philadelphia, Pennsylvania 19104-2688.

Library of Congress Cataloging-in-Publication Data

Stewart, G. W. (Gilbert W.)
    Afternotes on numerical analysis : a series of lectures on elementary numerical analysis presented at the University of Maryland at College Park and recorded after the fact / G.W. Stewart.
      p. cm.
    Includes bibliographical references and index.
    ISBN 0-89871-362-5 (pbk.)
    1. Numerical analysis   I. Title.
QA297.S785   1996
519.4--dc20
                                                          95-47768

siam. is a registered trademark.

Contents

Preface   ix

Nonlinear Equations   1

   Lecture 1 3
      By the dawn's early light 3
      Interval bisection 4
      Relative error 7
   Lecture 2 9
      Newton's method 9
      Reciprocals and square roots 11
      Local convergence analysis 12
      Slow death 14
   Lecture 3 17
      A quasi-Newton method 17
      Rates of convergence 20
      Iterating for a fixed point 21
      Multiple zeros 24
      Ending with a proposition 25
   Lecture 4 27
      The secant method 27
      Convergence 29
      Rate of convergence 31
      Multipoint methods 33
      Muller's method 33
      The linear-fractional method 34
   Lecture 5 37
      A hybrid method 37
      Errors, accuracy, and condition numbers 40

Floating-Point Arithmetic   43

   Lecture 6 45
      Floating-point numbers 45
      Overflow and underflow 47
      Rounding error 48
      Floating-point arithmetic 49
   Lecture 7 53
      Computing sums 53
      Backward error analysis 55
      Perturbation analysis 57
      Cheap and chippy chopping 58
   Lecture 8 61

	Cancellation	61
	The quadratic equation	61
	That fatal bit of rounding error	63
	Envoi	65

Linear Equations   67

Lecture 9 . . . 69
    Matrices, vectors, and scalars . . . 69
    Operations with matrices . . . 70
    Rank-one matrices . . . 73
    Partitioned matrices . . . 74

Lecture 10 . . . 77
    The theory of linear systems . . . 77
    Computational generalities . . . 78
    Triangular systems . . . 79
    Operation counts . . . 81

Lecture 11 . . . 83
    Memory considerations . . . 83
    Row-oriented algorithms . . . 83
    A column-oriented algorithm . . . 84
    General observations on row and column orientation . . . 86
    Basic linear algebra subprograms . . . 86

Lecture 12 . . . 89
    Positive-definite matrices . . . 89
    The Cholesky decomposition . . . 90
    Economics . . . 94

Lecture 13 . . . 97
    Inner-product form of the Cholesky algorithm . . . 97
    Gaussian elimination . . . 98

Lecture 14 . . . 103
    Pivoting . . . 103
    BLAS . . . 108
    Upper Hessenberg and tridiagonal systems . . . 110

Lecture 15 . . . 113
    Vector norms . . . 113
    Matrix norms . . . 114
    Relative error . . . 115
    Sensitivity of linear systems . . . 116

Lecture 16 . . . 119
    The condition of a linear system . . . 119
    Artificial ill-conditioning . . . 120
    Rounding error and Gaussian elimination . . . 122
    Comments on the error analysis . . . 125

Lecture 17 127
    Introduction to a project 127
    More on norms 127
    The wonderful residual 128
    Matrices with known condition numbers 129
    Invert and multiply 130
    Cramer's rule 130
    Submission 131

Polynomial Interpolation   **133**

Lecture 18 135
    Quadratic interpolation 135
    Shifting 136
    Polynomial interpolation 137
    Lagrange polynomials and existence 137
    Uniqueness 138

Lecture 19 141
    Synthetic division 141
    The Newton form of the interpolant 142
    Evaluation 142
    Existence and uniqueness 143
    Divided differences 144

Lecture 20 147
    Error in interpolation 147
    Error bounds 149
    Convergence 150
    Chebyshev points 151

Numerical Integration   **155**

Lecture 21 157
    Numerical integration 157
    Change of intervals 158
    The trapezoidal rule 158
    The composite trapezoidal rule 160
    Newton–Cotes formulas 161
    Undetermined coefficients and Simpson's rule 162

Lecture 22 165
    The Composite Simpson rule 165
    Errors in Simpson's rule 166
    Treatment of singularities 167
    Gaussian quadrature: The idea 169

Lecture 23 171
    Gaussian quadrature: The setting 171

Orthogonal polynomials 171
Existence 173
Zeros of orthogonal polynomials 174
Gaussian quadrature 175
Error and convergence 176
Examples 176

Numerical Differentiation   **179**
   Lecture 24 181
      Numerical differentiation and integration 181
      Formulas from power series 182
      Limitations 184

Bibliography   **187**
   Introduction 187
   References 187

Index   **191**

Preface

In the spring of 1993, I took my turn at teaching our upper-division course in introductory numerical analysis. The topics covered were nonlinear equations, computer arithmetic, linear equations, polynomial interpolation, numerical integration, and numerical differentiation. The continuation of the course is a sequence of two graduate courses, which offer a selection of complementary topics. I taught Tuesday-Thursday classes of eighty-five minutes.

The textbook was *Numerical Analysis* by David Kincaid and Ward Cheney. However, I usually treat textbooks as supplemental references and seldom look at them while I am preparing lectures. The practice has the advantage of giving students two views of the subject. But in the end all I have is a handful of sketchy notes and vague recollections of what I said in class.

To find out what I was actually teaching I decided to write down each lecture immediately after it was given while it was still fresh in my mind. I call the results afternotes. If I had known what I was letting myself in for, I would have never undertaken the job. Writing to any kind of deadline is difficult; writing to a self-imposed deadline is torture. Yet now I'm glad I did it. I learned a lot about the anatomy of a numerical analysis course.

I also had an ulterior motive. Most numerical analysis books, through no fault of their authors, are a bit ponderous. The reason is they serve too many masters. They must instruct the student (and sometimes the teacher). They must also contain enough topics to allow the instructor to select his or her favorites. In addition, many authors feel that their books should be references that students can take with them into the real world. Now there are various ways to combine these functions in a single book — and they all slow down the exposition. In writing these afternotes, I was curious to see if I could give the subject some narrative drive by screwing down on the focus. You will have to judge how well I have succeeded.

So what you have here is a replica of what I said in class. Not a slavish replica. The blackboard and the printed page are different instruments and must be handled accordingly. I corrected errors, big and small, whenever I found them. Moreover, when I saw better ways of explaining things, I did not hesitated to rework what I had originally presented. Still, the correspondence is close, and each section of the notes represents about a class period's worth of talking.

In making these notes available, I hope that they will be a useful supplement for people taking a numerical course or studying a conventional textbook on their own. They may also be a source of ideas for someone teaching numerical analysis for the first time. To increase their utility I have appended a brief bibliography.

The notes were originally distributed over the Internet, and they have benefited from the feedback. I would like to thank John Carroll, Bob Funderlic,

David Goldberg, Murli Gupta, Nick Higham, Walter Hoffman, Keith Lindsay, and Dean Schulze for their comments. I am also indebted to the people at SIAM who saw the notes through production: to Vickie Kearn for resurrecting them from the oblivion of my class directory, to Jean Anderson for a painstaking job of copy editing, and to Corey Gray for an elegant design.

Above all I owe one to my wife, Astrid Schmidt-Nielsen, who was a patient and encouraging workstation widow throughout the writing of these notes. They are dedicated to her.

G. W. Stewart
College Park, MD

Nonlinear Equations

Lecture 1

Nonlinear Equations

By the Dawn's Early Light
Interval Bisection
Relative Error

By the dawn's early light

1. For a simple example of a nonlinear equation, consider the problem of aiming a cannon to hit a target at distance d. The cannon is assumed to have muzzle velocity V_0 and elevation θ.

To determine how far the cannon ball travels, note that the vertical component of the muzzle velocity is $V_0 \sin \theta$. Since the ball is moving vertically against the acceleration of gravity g, its vertical position $y(t)$ satisfies the differential equation

$$y''(t) = -g, \qquad \begin{cases} y(0) = 0, \\ y'(0) = V_0 \sin \theta. \end{cases}$$

The solution is easily seen to be

$$y(t) = V_0 t \sin \theta - \frac{1}{2} g t^2.$$

Thus the ball will hit the ground at time

$$T = \frac{2 V_0 \sin \theta}{g}.$$

Since the horizontal component of the velocity is $V_0 \cos \theta$, the ball will travel a distance of $T V_0 \cos \theta$. Thus to find the elevation we have to solve the equation

$$\frac{2 V_0^2 \cos \theta \sin \theta}{g} = d,$$

or equivalently

$$f(\theta) \equiv \frac{2 V_0^2 \cos \theta \sin \theta}{g} - d = 0. \tag{1.1}$$

2. Equation (1.1) exhibits a number of features associated with nonlinear equations. Here is a list.

• The equation is an idealization. For example, it does not take into account the resistance of air. Again, the derivation assumes that the muzzle of the cannon is level with the ground—something that is obviously not true. The lesson is that when you are presented with a numerical problem in the abstract it is usually a good idea to ask about where it came from before working hard to solve it.

- The equation may not have a solution. Since $\cos\theta\sin\theta$ assumes a maximum of $\frac{1}{2}$ at $\theta = \frac{\pi}{4}$, there will be no solution if

$$d > \frac{V_0^2}{g}.$$

- Solutions, when they exist, are not unique. If there is one solution, then there are infinitely many, since sin and cos are periodic. These solutions represent a rotation of the cannon elevation through a full circle. Any resolution of the problem has to take these spurious solutions into account.

- If $d < V_0^2/4g$, and $\theta_* < \frac{\pi}{2}$ is a solution, then $\frac{\pi}{2} - \theta_*$ is also a solution. Both solutions are meaningful, but as far as the gunner is concerned, one may be preferable to the other. You should find out which.

- The function f is simple enough to be differentiated. Hence we can use a method like Newton's method.

- In fact, (1.1) can be solved directly. Just use the relation $2\sin\theta\cos\theta = \sin 2\theta$. It is rare for things to turn out this nicely, but you should try to simplify before looking for numerical solutions.

- If we make the model more realistic, say by including air resistance, we may end up with a set of differential equations that can only be solved numerically. In this case, analytic derivatives will not be available, and one must use a method that does not require derivatives, such as a quasi-Newton method (§3.1).

Interval bisection

3. In practice, a gunner may determine the range by trial and error, raising and lowering the cannon until the target is obliterated. The numerical analogue of this process is interval bisection. From here on we will consider the general problem of solving the equation

$$f(x) = 0. \tag{1.2}$$

4. The theorem underlying the bisection method is called the *intermediate value theorem*.

> If f is continuous on $[a, b]$ and g lies between $f(a)$ and $f(b)$, then there is a point $x \in [a, b]$ such that $g = f(x)$.

1. Nonlinear Equations

5. The intermediate value theorem can be used to establish the existence of a solution of (1.2). Specifically if $\text{sign}[f(a)] \neq \text{sign}[f(b)]$, then zero lies between $f(a)$ and $f(b)$. Consequently, there is a point x in $[a, b]$ such that $f(x) = 0$.

6. We can turn this observation into an algorithm that brackets the root in intervals of ever decreasing width. Specifically, suppose that $\text{sign}[f(a)] \neq \text{sign}[f(b)]$, and to avoid degeneracies suppose that neither $f(a)$ nor $f(b)$ is zero. In this case, we will call the interval $[a, b]$ a *nontrivial bracket* for a root of f. Here, and always throughout these notes, $[a, b]$ will denote the set of points between a and b, inclusive, with no implication that $a \leq b$.

Now let $c = \frac{a+b}{2}$. There are three possibilities.

1. $f(c) = 0$. In this case we have found a solution of (1.2).
2. $f(c) \neq 0$ and $\text{sign}[f(c)] \neq \text{sign}[f(b)]$. In this case $[c, b]$ is a nontrivial bracket.
3. $f(c) \neq 0$ and $\text{sign}[f(a)] \neq \text{sign}[f(c)]$. In this case $[a, c]$ is a nontrivial bracket.

Thus we either solve the problem or we end up with a nontrivial bracket that is half the size of the original. The process can be repeated indefinitely, each repetition either solving the problem or reducing the length of the bracket by a factor of two. Figure 1.1 illustrates several iterations of this process. The numbers following the letters a and b indicate at which iteration these points were selected.

7. These considerations lead to the following algorithm. The input is a nontrivial bracket [a, b] and the function values fa and fb at the endpoints. In addition we need a stopping criterion eps ≥ 0. The algorithm usually returns a bracket of length not greater than eps. (For the exception see §1.9.)

```
while (abs(b-a) > eps){
    c = (b+a)/2;
    if (c==a || c==b)
        return;
    fc = f(c);
    if (fc == 0){
        a = b = c;
        fa = fb = fc;
        return;
    }
    if (sign(fc) != sign(fb))
        {a = c; fa = fc;}
    else
        {b = c; fb = vc;}
}
return;
```
(1.3)

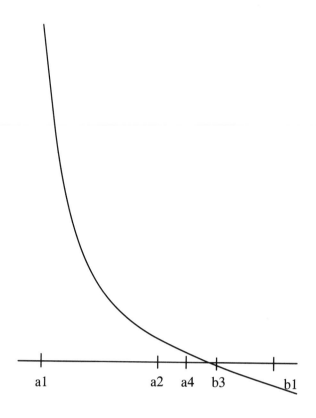

Figure 1.1. *Interval bisection.*

8. The hardest part about using the bisection algorithm is finding a bracket. Once it is found, the algorithm is guaranteed to converge, provided the function is continuous. Although later we shall encounter algorithms that converge much faster, the bisection method converges steadily. If $L_0 = |b - a|$ is the length of the original bracket, after k iterations the bracket has length

$$L_k = \frac{L_0}{2^k}.$$

Since the algorithm will stop when $L_k \leq$ eps, it will require

$$\left\lceil \log_2 \frac{L_0}{\text{eps}} \right\rceil$$

iterations to converge. Thus if $L_0 = 1$ and eps $= 10^{-6}$, the iteration will require 20 iterations.

9. The statement

1. Nonlinear Equations

```
       if (c==a || c==b)
           return;
```

is a concession to the effects of rounding error. If eps is too small, it is possible for the algorithm to arrive at the point where (a+b)/2 evaluates to either a or b, after which the algorithm will loop indefinitely. In this case the algorithm, having given its all, simply returns.[1]

Relative error

10. The convergence criterion used in (1.3) is based on *absolute error*; that is, it measures the error in the result without regard to the size of the result. This may or may not be satisfactory. For example, if eps $= 10^{-6}$ and the zero in question is approximately one, then the bisection routine will return roughly six accurate digits. However, if the root is approximately 10^{-7}, we can expect no figures of accuracy: the final bracket can actually contain zero.

11. If a certain number of significant digits are required, then a better measure of error is *relative error*. Formally, if y is an approximation to $x \neq 0$, then the relative error in y is the number

$$\rho = \frac{|y-x|}{|x|}.$$

Alternatively, y has relative error ρ, if there is a number ϵ with $|\epsilon| = \rho$ such that

$$y = x(1+\epsilon).$$

12. The following table of approximations to $e = 2.7182818\ldots$ illustrates the relation of relative error and significant digits.

Approximation	ρ
2.	$2 \cdot 10^{-1}$
2.7	$6 \cdot 10^{-3}$
2.71	$3 \cdot 10^{-3}$
2.718	$1 \cdot 10^{-4}$
2.7182	$3 \cdot 10^{-5}$
2.71828	$6 \cdot 10^{-7}$

An examination of this table suggests the following.

> If x and y agree to k decimal digits, then the relative error in y will be approximately 10^{-k}.

---

[1] Thanks to Urs von Matt for pointing this out.

13. If we exclude tricky cases like $x = 2.0000$ and $y = 1.9999$, in which the notion of agreement of significant digits is not well defined, the relation between agreement and relative error is not difficult to establish. Let us suppose, say, that x and y agree to six figures. Writing x above y, we have

$$x = X_1 X_2 X_3 X_4 X_5 X_6 X_7 X_8,$$
$$y = Y_1 Y_2 Y_3 Y_4 Y_5 Y_6 Y_7 Y_8.$$

Now since the digits X_7 and Y_7 must disagree, the smallest difference between x and y is obtained when, e.g.,

$$X_7 X_8 = 40,$$
$$Y_7 Y_8 = 38.$$

Thus $|x - y| \geq 2$, which is a lower bound on the difference. On the other hand, if X_7 is nine while Y_7 is zero, then $|y - x| < 10$, which is an upper bound. Thus

$$2 \leq |x - y| < 100.$$

Since $1 \leq X_1 \leq 9$, it is easy to see that

$$10^7 \leq |x| < 10^8.$$

Hence

$$0.2 \cdot 10^{-7} < \frac{|y - x|}{|x|} < 10^{-5};$$

i.e., the relative error is near 10^{-6}.

14. Returning to interval bisection, if we want a relative error of ρ in the answer, we might replace the convergence criterion in the **while** statement with

```
while(abs(b-a)/min(abs(a),abs(b)) > rho)
```

Note that this criterion is dangerous if the initial bracket straddles zero, since the quantity `min(abs(a),abs(b))` could approach zero.

Lecture 2

Nonlinear Equations

Newton's Method
Reciprocals and Square Roots
Local Convergence Analysis
Slow Death

Newton's method

1. Newton's method is an iterative method for solving the nonlinear equation

$$f(x) = 0. \qquad (2.1)$$

Like most iterative methods, it begins with a starting point x_0 and produces successive approximations $x_1, x_2, \ldots$. If x_0 is sufficiently near a root x_* of (2.1), the sequence of approximations will approach x_*. Usually the convergence is quite rapid, so that once the typical behavior of the method sets in, it requires only a few iterations to produce a very accurate approximation to the root. (The point x_* is also called a *zero* of the function f. The distinction is that equations have roots while functions have zeros.)

Newton's method can be derived in two ways: geometrically and analytically. Each has its advantages, and we will treat each in turn.

2. The geometric approach is illustrated in Figure 2.1. The idea is to draw a tangent to the curve $y = f(x)$ at the point $A = (x_0, f(x_0))$. The abscissa x_1 of the point $C = (x_1, 0)$ where the tangent intersects the axis is the new approximation. As the figure suggests, it will often be a better approximation to x_* than x_0.

To derive a formula for x_1, consider the distance $\overline{BC}$ from x_0 to x_1, which satisfies

$$\overline{BC} = \frac{\overline{BA}}{\tan \widehat{ACB}}.$$

But $\overline{BA} = f(x_0)$ and $\tan \widehat{ABC} = -f'(x_0)$ (remember the derivative is negative at x_0). Consequently,

$$x_1 = x_0 - \frac{f(x_0)}{f'(x_0)}.$$

If the iteration is carried out once more, the result is point D in Figure 2.1. In general, the iteration can be continued by defining

$$x_{k+1} = x_k - \frac{f(x_k)}{f'(x_k)}, \qquad k = 0, 1, \ldots.$$

9

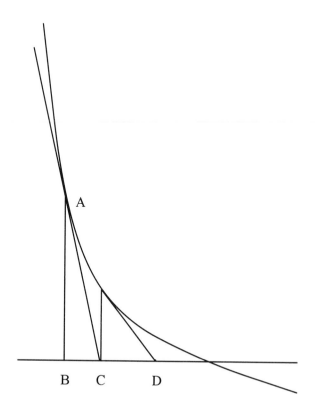

Figure 2.1. *Geometric illustration of Newton's method.*

3. The analytic derivation of Newton's method begins with the Taylor expansion

$$f(x) = f(x_0) + f'(x_0)(x - x_0) + \frac{1}{2}f''(\xi_0)(x - x_0)^2,$$

where as usual ξ_0 lies between x and x_0. Now if x_0 is near the zero x_* of f and $f'(x_0)$ is not too large, then the function

$$\hat{f}(x) = f(x_0) + f'(x_0)(x - x_0)$$

provides a good approximation to $f(x)$ in the neighborhood of x_*. For example, if $|f''(x)| \leq 1$ and $|x - x_0| \leq 10^{-2}$, then $|\hat{f}(x) - f(x)| \leq 10^{-4}$. In this case it is reasonable to assume that the solution of the equation $\hat{f}(x) = 0$ will provide a good approximation to x_*. But this solution is easily seen to be

$$x_1 = x_0 - \frac{f(x_0)}{f'(x_0)},$$

which is just the Newton iteration formula.

4. In some sense the geometric and analytic derivations of Newton's method say the same thing, since $y = \hat{f}(x)$ is just the equation of the tangent line $\overline{AC}$ in Figure 2.1. However, in other respects the approaches are complementary. For example, the geometric approach shows (informally) that Newton's method must converge for a function shaped like the graph in Figure 2.1 and a starting point for which the function is positive. On the other hand the analytic approach suggests fruitful generalizations. For example, if f' does not vary too much we might skip the evaluation of $f'(x_k)$ and iterate according to the formula

$$x_{k+1} = x_k - \frac{f(x_k)}{f'(x_0)}, \qquad k = 1, 2, \ldots. \qquad (2.2)$$

Reciprocals and square roots

5. If $a > 0$, the function

$$f(x) = \frac{1}{x} - a$$

has the single positive zero $x_* = a^{-1}$. If Newton's method is applied to this function, the result is the iteration

$$x_{k+1} = 2x_k - ax_k^2. \qquad (2.3)$$

In fact, the graph in Figure 2.1 is just a graph of $x^{-1} - 0.5$. Among other things, it shows that the iteration will converge from any starting point $x_0 > 0$ that is less than $\frac{1}{a}$. Because the iteration requires no divisions, it has been used as an alternative to hard-wired division in some computers.

6. If $a > 0$, the function

$$f(x) = x^2 - a$$

has the single positive zero $x_* = \sqrt{a}$. If Newton's method is applied to this function, the result is the iteration

$$x_{k+1} = \frac{1}{2}\left(x_k + \frac{a}{x_k}\right). \qquad (2.4)$$

This formula for approximating the square root was known to the Babylonians. Again, it is easy to see geometrically that if $x_0 > \sqrt{a}$ then the iteration converges to $\sqrt{a}$.

7. Newton's method does not have to converge. For example, if x_0 is too large in the iteration (2.3) for the reciprocal, then x_1 will be less than zero and the subsequent iterates will diverge to $-\infty$.

Local convergence analysis

8. We are going to show that if x_0 is sufficiently near a zero of x_* of f and
$$f'(x_*) \neq 0,$$
then Newton's method converges — ultimately with great rapidity. To simplify things, we will assume that f has derivatives of all orders. We will also set
$$\varphi(x) = x - \frac{f(x)}{f'(x)},$$
so that
$$x_{k+1} = \varphi(x_k).$$
The function φ is called the *iteration function* for Newton's method. Note that
$$\varphi(x_*) = x_* - \frac{f(x_*)}{f'(x_*)} = x_*.$$
Because x_* is unaltered by φ, it is called a *fixed point* of φ.

Finally we will set
$$e_k = x_k - x_*.$$
The quantity e_k is the error in x_k as an approximation to x_*. To say that $x_k \to x_*$ is the same as saying that $e_k \to 0$.

9. The local convergence analysis of Newton's method is typical of many convergence analyses. It proceeds in three steps.

   1. Obtain an expression for e_{k+1} in terms of e_k.
   2. Use the expression to show that $e_k \to 0$.
   3. Knowing that the iteration converges, assess how fast it converges.

10. The error formula can be derived as follows. Since $x_{k+1} = \varphi(x_k)$ and $x_* = \varphi(x_*)$,
$$e_{k+1} = x_{k+1} - x_* = \varphi(x_k) - \varphi(x_*).$$
By Taylor's theorem with remainder,
$$\varphi(x_k) - \varphi(x_*) = \varphi'(\xi_k)(x_k - x_*),$$
where ξ_k lies between x_k and x_*. It follows that
$$e_{k+1} = \varphi'(\xi_k) e_k. \qquad (2.5)$$
This is the error formula we need to prove convergence.

11. At first glance, the formula (2.5) appears difficult to work with since it depends on ξ_k which varies from iterate to iterate. However — and this is the

essence of any local convergence theorem — we may take our starting point close enough to x_* so that

$$|\varphi'(\xi_k)| \leq C < 1, \qquad k = 0, 1, \ldots.$$

Specifically, we have

$$\varphi'(x) = \frac{f(x)f''(x)}{f'(x)^2}.$$

Since $f(x_*) = 0$ and $f'(x_*) \neq 0$, it follows that $\varphi'(x_*) = 0$. Hence by continuity, there is an interval $I_\delta = [x_* - \delta, x_* + \delta]$ about x_* such that if $x \in I_\delta$ then

$$|\varphi'(x)| \leq C < 1.$$

Now suppose that $x_0 \in I_\delta$. Then since ξ_0 is between x_* and x_0, it follows that $\xi_0 \in I_\delta$. Hence from (2.5),

$$|e_1| \leq |\varphi''(\xi_k)||e_0| \leq C|e_0| \leq C\delta < \delta,$$

and $x_1 \in I_\delta$. Now since x_1 is in I_δ, so is x_2 by the same reasoning. Moreover,

$$|e_2| \leq C|e_1| \leq C^2|e_0|.$$

By induction, if $x_{k-1} \in I_\delta$, so is x_k, and

$$|e_k| \leq C|e_{k-1}| \leq C^k|e_0|.$$

Since $C^k \to 0$, it follows that $e_k \to 0$; that is, the sequence $x_0, x_1, \ldots$ converges to x_*, which is what we wanted to show.

12. To assess the rate of convergence, we turn to a higher-order Taylor expansion. Since $\varphi'(x_*) = 0$, we have

$$\varphi(x_k) - \varphi(x_*) = \frac{1}{2}\varphi''(\eta_k)(x_k - x_*)^2,$$

where η_k lies between x_k and x_*. Hence

$$e_{k+1} = \frac{1}{2}\varphi''(\eta_k)e_k^2.$$

Since η_k approaches x_* along with x_k, it follows that

$$\lim_{k \to \infty} \frac{e_{k+1}}{e_k^2} = \frac{1}{2}\varphi''(x_*) \equiv \frac{f''(x_*)}{2f'(x_*)}.$$

In other words,

$$e_{k+1} \cong \frac{f''(x_*)}{2f'(x_*)} e_k^2. \qquad (2.6)$$

A sequence whose errors behave like this is said to be *quadratically convergent*.

13. To see informally what quadratic convergence means, suppose that the multiplier of e_k^2 in (2.6) is one and that $e_0 = 10^{-1}$. Then $e_1 \cong 10^{-2}$, $e_3 \cong 10^{-4}$, $e_4 \cong 10^{-8}$, $e_5 \cong 10^{-16}$, and so on. Thus if x_* is about one in magnitude, the first iterate is accurate to about two places, the second to four, the third to eight, the fourth to sixteen, and so on. In this case each iteration of Newton's method doubles the number of accurate figures.

For example, if the formula (2.4) is used to approximate the square root of ten, starting from three, the result is the following sequence of iterates.

$$3.$$
$$3.16$$
$$3.1622$$
$$3.16227766016$$
$$3.16227766016838$$

Only the correct figures are displayed, and they roughly double at each iteration. The last iteration is exceptional, because the computer I used carries only about fifteen decimal digits.

14. For a more formal analysis, recall that the number of significant figures in an approximation is roughly the negative logarithm of the relative error (see §1.12). Assume that $x_* \neq 0$, and let ρ_k denote the relative error in x_k. Then from (2.6) we have

$$\rho_{k+1} \cong \frac{|x_* f''(x_*)|}{2|f'(x_*)|}\rho_k^2 \equiv K\rho_k^2.$$

Hence

$$-\log \rho_{k+1} \cong -2\log \rho_k - \log K.$$

As the iteration converges, $-\log \rho_k \to \infty$, and it overwhelms the value of $\log K$. Hence

$$-\log \rho_{k+1} \cong -2\log \rho_k,$$

which says that x_{k+1} has twice as many significant figures as x_k.

Slow death

15. The convergence analysis we have just given shows that if Newton's method converges to a zero x_* for which $f'(x_*) \neq 0$ then in the long run it must converge quadratically. But the run can be very long indeed.

For example, in §2.5 we noted that the iteration

$$x_{k+1} = 2x_k - ax_k^2$$

will converge to a^{-1} starting from any point less than a^{-1}. In particular, if $a < 1$, we can take a itself as the starting value.

But suppose that $a = 10^{-10}$. Then

$$x_1 = 2 \cdot 10^{-10} + 10^{-30} \cong 2 \cdot 10^{-10}.$$

Thus for practical purposes the first iterate is only twice the size of the starting value. Similarly, the second iterate will be about twice the size of the first. This process of doubling the sizes of the iterates continues until $x_k \cong 10^{10}$, at which point quadratic convergence sets in. Thus we must have $2^k \cdot 10^{-10} \cong 10^{10}$ or $k \cong 66$ before we begin to see quadratic convergence. That is a lot of work to compute the reciprocal of a number.

16. All this does not mean that the iteration is bad, just that it needs a good starting value. Sometimes such a value is easy to obtain. For example, suppose that $a = f \cdot 2^e$, where $\frac{1}{2} \leq f < 1$ and we know e. These conditions are satisfied if a is represented as a binary floating-point number on a computer. Then $a^{-1} = f^{-1} \cdot 2^{-e}$. Since $1 < f^{-1} \leq 2$, the number $2^{-e} < a^{-1}$ provides a good starting value.

Lecture 3

Nonlinear Equations

A Quasi-Newton Method
Rates of Convergence
Iterating for a Fixed Point
Multiple Zeros
Ending with a Proposition

A quasi-Newton method

1. One of the drawbacks of Newton's method is that it requires the computation of the derivative $f'(x_k)$ at each iteration. There are three ways in which this can be a problem.

   1. The derivative may be very expensive to compute.
   2. The function f may be given by an elaborate formula, so that it is easy to make mistakes in differentiating f and writing code for the derivative.
   3. The value of the function f may be the result of a long numerical calculation. In this case the derivative will not be available as a formula.

2. One way of getting around this difficulty is to iterate according to the formula

$$x_{k+1} = x_k - \frac{f(x_k)}{g_k},$$

where g_k is an easily computed approximation to $f'(x_k)$. Such an iteration is called a *quasi-Newton* method.[2] There are many quasi-Newton methods, depending on how one approximates the derivative. For example, we will later examine the *secant method* in which the derivative is approximated by the difference quotient

$$g_k = \frac{f(x_k) - f(x_{k-1})}{x_k - x_{k-1}}. \tag{3.1}$$

Here we will analyze the simple case where g_k is constant, so that the iteration takes the form

$$x_{k+1} = x_k - \frac{f(x_k)}{g}. \tag{3.2}$$

We will call this method the *constant slope method*. In particular, we might take $g = f'(x_0)$, as in (2.2). Figure 3.1 illustrates the course of such an

---

[2] The term "quasi-Newton" usually refers to a class of methods for solving systems of simultaneous nonlinear equations.

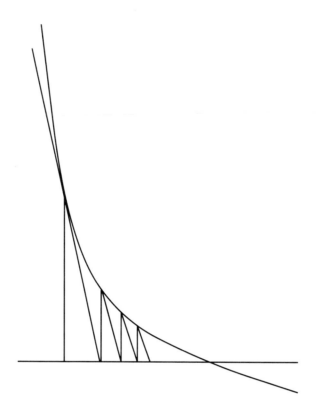

Figure 3.1. *The constant slope method.*

iteration.

3. Once again we have a local convergence theorem. Let

$$\varphi(x) = x - \frac{f(x)}{g}$$

be the iteration function and assume that

$$|\varphi'(x_*)| \equiv \left|1 - \frac{f'(x_*)}{g}\right| < 1. \tag{3.3}$$

Arguing as in §2.10, we can show that

$$e_{k+1} = \varphi'(\xi_k)e_k, \tag{3.4}$$

where ξ_k lies between x_k and x_*. Since $|\varphi'(x_*)| < 1$, there is an interval I about x_* such that

$$|\varphi(x)| \leq C < 1 \quad \text{whenever} \quad x \in I.$$

3. Nonlinear Equations

Arguing as in §2.11, we find that if $x_0 \in I$ then

$$|e_k| \leq C^k |e_0|,$$

which implies that the iteration converges.

4. To assess the quality of the convergence, note that (3.4) implies that

$$\lim_{k \to \infty} \frac{e_{k+1}}{e_k} = \varphi'(x_*),$$

or asymptotically

$$e_{k+1} \cong \varphi'(x_*) e_k.$$

Thus if $\varphi'(x_*) \neq 0$, each iteration reduces the error by roughly a factor of $\varphi'(x_*)$. Such convergence is called *linear convergence*.

5. It is worth noting that the convergence proof for the constant slope method (3.2) is the same as the proof for Newton's method itself — with one important exception. The iteration function for Newton's method satisfies $\varphi'(x_*) = 0$, from which it follows that the constant $C < 1$ exists regardless of the function f (provided that $f'(x_*) \neq 0$). For the quasi-Newton method, we must *postulate* that $|\varphi'(x_*)| < 1$ in order to insure the existence of the constant $C < 1$. This difference is a consequence of the fact that the denominator g in the quasi-Newton method is a free parameter, which, improperly chosen, can cause the method to fail.

6. The methods also differ in their rates of convergence: Newton's method converges quadratically, while the constant slope method converges linearly (except in the unlikely case where we have chosen $g = f'(x_*)$, so that $\varphi'(x_*) = 0$). Now in some sense all quadratic convergence is the same. Once it sets in, it doubles the number of significant figures at each step. Linear convergence is quite different. Its speed depends on the ratio

$$\rho = \lim_{k \to \infty} \frac{x_{k+1} - x_*}{x_k - x_*}.$$

If ρ is near one, the convergence will be slow. If it is near zero, the convergence will be fast.

7. It is instructive to consider the case where

$$x_{k+1} - x_* = \rho(x_k - x_*),$$

so that the error is reduced exactly by a factor of ρ at each iteration. In this case

$$x_k - x_* = \rho^k (x_0 - x_*).$$

It follows that to reduce the error by a factor of ϵ, we must have $\rho^k \leq \epsilon$ or

$$k = \left\lceil \frac{\log \epsilon}{\log \rho} \right\rceil.$$

The following table gives values of k for representative values of ρ and ϵ.

		ρ				
		.99	.90	.50	.10	.01
	10^{-5}	1146	110	17	6	3
ϵ	10^{-10}	2292	219	34	11	6
	10^{-15}	3437	328	50	16	8

The numbers show that convergence can range from quite fast to very slow. With $\rho = 0.01$ the convergence is not much worse than quadratic convergence — at least in the range of reductions we are considering here. When $\rho = 0.99$ the convergence is very slow. Note that such rates, and even slower ones, arise in real life. The fact that some algorithms creep toward a solution is one of the things that keeps supercomputer manufacturers in business.

Rates of convergence

8. We are going to derive a general theory of iteration functions. However, first we must say something about rates of convergence. Here we will assume that we have a sequence $x_0, x_1, x_2, \ldots$ converging to a limit x_*.

9. If there is a constant ρ satisfying $0 < |\rho| < 1$ such that

$$\lim_{k \to \infty} \frac{x_{k+1} - x_*}{x_k - x_*} = \rho, \qquad (3.5)$$

then the sequence $\{x_k\}$ is said to converge *linearly with ratio (or rate)* ρ. We have already treated linear convergence above.[3]

10. If the ratio in (3.5) converges to zero, the convergence is said to be *superlinear*. Certain types of superlinear convergence can be characterized as follows. If there is a number $p > 1$ and a positive constant C such that

$$\lim_{k \to \infty} \frac{|x_{k+1} - x_*|}{|x_k - x_*|^p} = C,$$

then the sequence is said to *converge with order* p. When $p = 2$ the convergence is quadratic. When $p = 3$ the convergence is cubic. In general, the analysis of quadratic convergence in §2.14 can be adapted to show that the number of correct figures in a sequence exhibiting pth order convergence increases by a

---

[3] If $\rho = 1$, the convergence is sometimes called *sublinear*. The sequence $\{\frac{1}{k}\}$ converges sublinearly to zero.

factor of about p from iteration to iteration. Note that p does not have to be an integer. Later we shall see the secant method (3.1) typically converges with order $p = 1.62\ldots$.[4]

11. You will not ordinarily encounter rates of convergence greater than cubic, and even cubic convergence occurs only in a few specialized algorithms. There are two reasons. First, the extra work required to get higher-order convergence may not be worth it — especially in finite precision, where the accuracy that can be achieved is limited. Second, higher-order methods are often less easy to apply. They generally require higher-order derivatives and more accurate starting values.

Iterating for a fixed point

12. The essential identity of the local convergence proof for Newton's method and the quasi-Newton method suggests that they both might be subsumed under a general theory. Here we will develop such a theory. Instead of beginning with an equation of the form $f(x) = 0$, we will start with a function φ having a *fixed point* x_* — that is, a point x_* for which $\varphi(x_*) = x_*$ — and ask when the iteration

$$x_{k+1} = \varphi(x_k), \qquad k = 0, 1, \ldots \tag{3.6}$$

converges to x_*. This iterative method for finding a fixed point is called *the method of successive substitutions*.

13. The iteration (3.6) has a useful geometric interpretation, which is illustrated in Figures 3.2 and 3.3. The fixed point x_* is the abscissa of the intersection of the graph of $\varphi(x)$ with the line $y = x$. The ordinate of the function $\varphi(x)$ at x_0 is the value of x_1. To turn this ordinate into an abscissa, reflect it in the line $y = x$. We may repeat this process to get x_2, x_3, and so on. It is seen that the iterates in Figure 3.2 zigzag into the fixed point, while in Figure 3.3 they zigzag away: the one iteration converges if you start near enough to the fixed point, whereas the other diverges no matter how close you start. The fixed point in the first example is said to be *attractive,* and the one in the second example is said to be *repulsive*.

14. It is the value of the derivative of φ at the fixed point that makes the difference in these two examples. In the first the absolute value of the derivative is less than one, while in the second it is greater than one. (The derivatives here are both positive. It is instructive to draw iteration graphs in which the derivatives at the fixed point are negative.) These examples along with our earlier convergence proofs suggest that what is necessary for a method of successive substitutions to converge is that the absolute value of the derivative be less than one at the fixed point. Specifically, we have the following result.

---

[4]It is also possible for a sequence to converge superlinearly but not with order $p > 1$. The sequence $\frac{1}{k!}$ is an example.

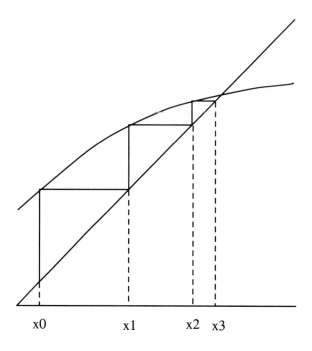

x0   x1   x2 x3

Figure 3.2. *An attractive fixed point.*

> If
> $$|\varphi'(x_*)| < 1,$$
> then there is an interval $I_\delta = [x_* - \delta, x_* + \delta]$ such that the iteration (3.6) converges to x_* whenever $x_0 \in I_\delta$. If $\varphi'(x_*) \neq 0$, then the convergence is linear with ratio $\varphi'(x_*)$. On the other hand, if
> $$0 = \varphi'(x_*) = \varphi''(x_*) = \cdots = \varphi^{(p-1)}(x_*) \neq \varphi^{(p)}(x_*), \qquad (3.7)$$
> then the convergence is of order p.

15. We have essentially seen the proof twice over. Convergence is established exactly as for Newton's method or the constant slope method. Linear convergence in the case where $\varphi'(x_*) \neq 0$ is verified as it was for the constant slope method. For the case where (3.7) holds, we need to verify that the convergence is of order p. In the usual notation, by Taylor's theorem

$$e_{k+1} = \frac{1}{p!}\varphi^{(p)}(\xi_k)e_k^p.$$

Since $\xi_k \to x_*$, it follows that

$$\lim_{k \to \infty} \frac{e_{k+1}}{e_k^p} = \frac{1}{p!}\varphi^{(p)}(x_*) \neq 0,$$

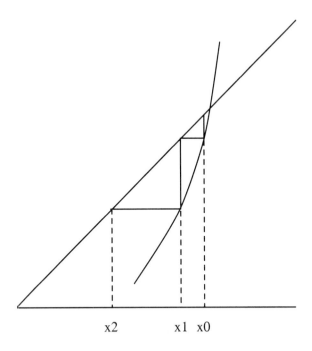

Figure 3.3. *A repulsive fixed point.*

which establishes the pth-order convergence.

16. Armed with this result, we can return to Newton's method and the constant slope method. For Newton's method we have

$$\varphi'(x_*) = \frac{f(x_*)f''(x_*)}{f'(x_*)^2} = 0$$

(remember that $f'(x_*)$ is assumed to be nonzero). Thus Newton's method is seen to be *at least* quadratically convergent. Since

$$\varphi''(x_*) = \frac{f''(x_*)}{f'(x_*)},$$

Newton's method will converge faster than quadratically only when $f''(x_*) = 0$.

For the constant slope method we have

$$\varphi'(x_*) = 1 - \frac{f'(x_*)}{g}.$$

The requirement that the absolute value of this number be less than one is precisely the condition (3.3).

Multiple zeros

17. Up to now we have considered only a *simple zero* of the function f, that is, a zero for which $f'(x_*) \neq 0$. We will now consider the case where

$$0 = f'(x_*) = f''(x_*) = \cdots = f^{(m-1)}(x_*) \neq f^{(m)}(x_*).$$

By Taylor's theorem

$$f(x) = (x - x_*)^m \frac{f^{(m)}(\xi_x)}{m!},$$

where ξ_x lies between x_* and x. If we set $g(x) = f^{(m)}(\xi_x)/m!$, then

$$f(x) = (x - x_*)^m g(x), \tag{3.8}$$

where g is continuous at x_* and $g(x_*) \neq 0$. Thus, when x is near x_*, the function $f(x)$ behaves like a polynomial with a zero of multiplicity m at x_*. For this reason we say that x_* is a *zero of multiplicity m of f*.

18. We are going to use the fixed-point theory developed above to assess the behavior of Newton's method at a multiple root. It will be most convenient to use the form (3.8). We will assume that g is twice differentiable.

19. Since $f'(x) = m(x - x_*)^{m-1} g(x) + (x - x_*)^m g'(x)$, the Newton iteration function for f is

$$\varphi(x) = x - \frac{(x - x_*)^m g(x)}{m(x - x_*)^{m-1} g(x) + (x - x_*)^m g'(x)} = x - \frac{(x - x_*)g(x)}{mg(x) - (x - x_*)g'(x)}.$$

From this we see that φ is well defined at x_* and

$$\varphi(x_*) = x_*.$$

According to fixed-point theory, we have only to evaluate the derivative of φ at x_* to determine if x_* is an attractive fixed point. We will skip the slightly tedious differentiation and get straight to the result:

$$\varphi'(x_*) = 1 - \frac{1}{m}.$$

Therefore, Newton's method converges to a multiple zero from any sufficiently close approximation, and the convergence is linear with ratio $1 - \frac{1}{m}$. In particular for a double root, the ratio is $\frac{1}{2}$, which is comparable with the convergence of interval bisection.

Ending with a proposition

20. Although roots that are exactly multiple are not common in practice, the above theory says something about how Newton's method behaves with a nearly multiple root. An amusing example is the following proposition for computing a zero of a polynomial of odd degree.

Let
$$f(x) = x^n + a_{n-1}x^{n-1} + \cdots + a_0, \qquad (3.9)$$

where n is odd. Since $f(x) > 0$ for large positive x and $f(x) < 0$ for large negative x, by the intermediate value theorem f has a real zero. Moreover, you can see graphically that if x_0 is greater than the largest zero of f, then Newton's method converges from x_0. The proposition, then, is to choose a very large value of x_0 (there are ways of choosing x_0 to be greater than the largest root), and let Newton's method do its thing.

The trouble with this proposition is that if x is very large, the term x^n in (3.9) dominates the others, and $f(x) \cong x^n$. In other words, from far out on the x-axis, f appears to have a zero of multiplicity n at zero. If Newton's method is applied, the error in each iterate will be reduced by a factor of only $1 - \frac{1}{n}$ (when $n = 100$, this is a painfully slow 0.99). In the long run, the iterates will arrive near a zero, after which quadratic convergence will set in. But, as we have had occasion to observe, in the long run we are all dead.

Lecture 4

Nonlinear Equations

The Secant Method
Convergence
Rate of Convergence
Multipoint Methods
Muller's Method
The Linear-Fractional Method

The secant method

1. Recall that in a quasi-Newton method we iterate according to the formula

$$x_{k+1} = x_k - \frac{f(x_k)}{g_k}, \tag{4.1}$$

where the numbers g_k are chosen to approximate $f'(x_k)$. One way of calculating such an approximation is to choose a step size h_k and approximate $f'(x_k)$ by the difference quotient

$$g_k = \frac{f(x_k + h_k) - f(x_k)}{h_k}.$$

There are two problems with this approach.

First, we have to determine the numbers h_k. If they are too large, the approximations to the derivative will be inaccurate and convergence will be retarded. If they are too small, the derivatives will be inaccurate owing to rounding error (we will return to this point in Lecture 24).

Second, the procedure requires one extra function evaluation per iteration. This is a serious problem if function evaluations are expensive.

2. The key to the secant method is to observe that once the iteration is started we have two nearby points, x_k and x_{k-1}, where the function has been evaluated. This suggests that we approximate the derivative by

$$g_k = \frac{f(x_k) - f(x_{k-1})}{x_k - x_{k-1}}.$$

The iteration (4.1) then takes the form

$$x_{k+1} = x_k - \frac{f(x_k)(x_k - x_{k-1})}{f(x_k) - f(x_{k-1})} = \frac{x_{k-1} f(x_k) - x_k f(x_{k-1})}{f(x_k) - f(x_{k-1})}. \tag{4.2}$$

This iteration is called the *secant method*.

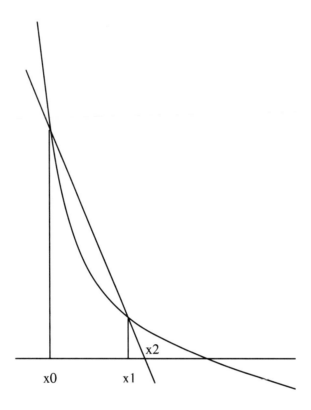

Figure 4.1. *The secant method.*

3. The secant method derives its name from the following geometric interpretation of the iteration. Given x_0 and x_1, draw the secant line through the graph of f at the points $(x_0, f(x_0))$ and $(x_1, f(x_1))$. The point x_2 is the abscissa of the intersection of the secant line with the x-axis. Figure 4.1 illustrates this procedure. As usual, a graph of this kind can tell us a lot about the convergence of the method in particular cases.

4. If we set

$$\varphi(u,v) = u - \frac{f(u)(u-v)}{f(u) - f(v)} = \frac{vf(u) - uf(v)}{f(u) - f(v)}, \qquad (4.3)$$

then the iteration (4.2) can be written in the form

$$x_{k+1} = \varphi(x_k, x_{k-1}).$$

Thus φ plays the role of an iteration function. However, because it has two arguments, the secant method is called a *two-point method*.

5. Although φ is indeterminate for $u = v$, we may remove the indeterminacy

4. Nonlinear Equations

by setting
$$\varphi(u,u) = u - \frac{f(u)}{f'(u)}.$$

In other words, the secant method reduces to Newton's method in the confluent case where $x_k = x_{k-1}$. In particular, it follows that

$$\varphi(x_*, x_*) = x_*,$$

so that x_* is a fixed point of the iteration.

Convergence

6. Because the secant method is a two-point method, the fixed-point theory developed above does not apply. In fact, the convergence analysis is considerably more complicated. But it still proceeds in the three steps outlined in §2.9: (1) find a recursion for the error, (2) show that the iteration converges, and (3) assess the rate of convergence. Here we will consider the first two steps.

7. It is a surprising fact that we do not need to know the specific form (4.3) of the iteration function to derive an error recurrence. Instead we simply use the fact that if we input the answer we get the answer back. More precisely, if one of the arguments of φ is the zero x_* of f, then φ returns x_*; i.e.,

$$\varphi(u, x_*) \equiv x_* \quad \text{and} \quad \varphi(x_*, v) \equiv x_*.$$

Since $\varphi(u, x_*)$ and $\varphi(x_*, v)$ are constant, their derivatives with respect to u and v are zero:

$$\varphi_u(u, x_*) \equiv 0 \quad \text{and} \quad \varphi_v(x_*, v) \equiv 0.$$

The same is true of the second derivatives:

$$\varphi_{uu}(u, x_*) \equiv 0 \quad \text{and} \quad \varphi_{vv}(x_*, v) \equiv 0.$$

8. To get an error recursion, we begin by expanding φ about (x_*, x_*) in a two-dimensional Taylor series. Specifically,

$$\begin{aligned}\varphi(x_* + p, x_* + q) &= \varphi(x_*, x_*) + \varphi_u(x_*, x_*)p + \varphi_v(x_*, x_*)q \\ &+ \tfrac{1}{2}[\varphi_{uu}(x_* + \theta p, x_* + \theta q)p^2 \\ &+ 2\varphi_{uv}(x_* + \theta p, x_* + \theta q)pq + \varphi_{vv}(x_* + \theta p, x_* + \theta q)q^2],\end{aligned}$$

where $\theta \in [0,1]$. Since $\varphi(x_*, x_*) = x_*$ and $\varphi_u(x_*, x_*) = \varphi_v(x_*, x_*) = 0$,

$$\begin{aligned}\varphi(x_* + p, x_* + q) &= x_* + \tfrac{1}{2}[\varphi_{uu}(x_* + \theta p, x_* + \theta q)p^2 \\ &+ 2\varphi_{uv}(x_* + \theta p, x_* + \theta q)pq + \varphi_{vv}(x_* + \theta p, x_* + \theta q)q^2].\end{aligned} \quad (4.4)$$

The term containing the cross product pq is just what we want, but the terms in p^2 and q^2 require some massaging. Since $\varphi_{uu}(x_* + \theta p, x_*) = 0$, it follows from a Taylor expansion in the second argument that

$$\varphi_{uu}(x_* + \theta p, x_* + \theta q) = \theta \varphi_{uuv}(x_* + \theta p, x_* + \tau_q \theta q) q,$$

where $\tau_q \in [0, 1]$. Similarly,

$$\varphi_{vv}(x_* + \theta p, x_* + \theta q) = \theta \varphi_{uvv}(x_* + \tau_p \theta p, x_* + \theta q) p,$$

where $\tau_p \in [0, 1]$. Substituting these values in (4.4) gives

$$\begin{aligned}\varphi(x_* + p, x_* + q) = x_* + \frac{pq}{2}[\theta \varphi_{uuv}(x_* + \theta p, x_* + \tau_q \theta q) p \\ + 2\varphi_{uv}(x_* + \theta p, x_* + \theta q) + \theta \varphi_{uvv}(x_* + \tau_p \theta p, x_* + \theta q) q].\end{aligned} \quad (4.5)$$

9. Turning now to the iteration proper, let the starting values be x_0 and x_1, and let their errors be $e_0 = x_0 - x_*$ and $e_1 = x_1 - x_*$. Taking $p = e_1$ and $q = e_0$ in (4.5), we get

$$\begin{aligned}e_2 &= \varphi(x_* + e_1, x_* + e_0) - x_* \\ &= \frac{e_1 e_0}{2}[\theta \varphi_{uuv}(x_* + \theta e_1, x_* + \tau_{e_0} \theta e_0) e_1 \\ &\quad + 2\varphi_{uv}(x_* + \theta e_1, x_* + \theta e_0) + \theta \varphi_{uvv}(x_* + \tau_{e_1} \theta e_1, x_* + \theta e_0) e_0] \\ &\equiv \frac{e_1 e_0}{2} r(e_1, e_0).\end{aligned}$$
$$(4.6)$$

This is the error recurrence we need.

10. We are now ready to establish the convergence of the method. First note that
$$r(0,0) = 2\varphi_{uv}(x_*, x_*).$$
Hence there is a $\delta > 0$ such that if $|u|, |v| \leq \delta$ then

$$|vr(u,v)| \leq C < 1.$$

Now let $|e_0|, |e_1| \leq \delta$. From the error recurrence (4.6) it follows that $|e_2| \leq C|e_1| < |e_1| \leq \delta$. Hence
$$|e_1 r(e_2, e_1)| \leq C < 1,$$
and $|e_3| \leq C|e_2| \leq C^2|e_1|$. By induction

$$|e_k| \leq C^{k-1}|e_1|,$$

and since the right-hand side of this inequality converges to zero, we have $e_k \to 0$; i.e., the secant method converges from any two starting values whose errors are less than δ in absolute value.

4. Nonlinear Equations

Rate of convergence

11. We now turn to the convergence rate of the general two-point method. The first thing to note is that since

$$e_{k+1} = \frac{e_k e_{k-1}}{2} r(e_k, e_{k-1}) \qquad (4.7)$$

and $r(0,0) = 2\varphi_{uv}(x_*, x_*)$, we have

$$\lim_{k \to \infty} \frac{e_{k+1}}{e_k e_{k-1}} = \varphi_{uv}(x_*, x_*). \qquad (4.8)$$

If $\varphi_{uv}(x_*, x_*) \neq 0$, we shall say that the sequence $\{x_k\}$ exhibits *two-point convergence*.

12. We are going to show that two-point convergence is superlinear of order

$$p = \frac{1 + \sqrt{5}}{2} = 1.618\ldots.$$

This number is the largest root of the equation

$$p^2 - p - 1 = 0. \qquad (4.9)$$

Now there are two ways to establish this fact. The first is to derive (4.9) directly from (4.8), which is the usual approach. However, since we already know the value of p, we can instead set

$$s_k = \frac{|e_{k+1}|}{|e_k|^p} \qquad (4.10)$$

and use (4.8) to verify that the s_k have a nonzero limit.

13. You should be aware that some people object to this way of doing things because (they say) it hides the way the result — in this case the particular value of p — was derived. On the other hand, the mathematician Gauss is reported to have said that after you build a cathedral you don't leave the scaffolding around; and he certainly would have approved of the following proof. Both sides have good arguments to make; but as a practical matter, when you know or have guessed the solution of a problem it is often easier to verify that it works than to derive it from first principles.

14. From (4.10) we have

$$|e_k| = s_{k-1}|e_{k-1}|^p$$

and

$$|e_{k+1}| = s_k|e_k^p| = s_k s_{k-1}^p |e_{k-1}|^{p^2}.$$

From (4.7),

$$|r_k| \equiv |r(e_k, e_{k-1})| = \frac{s_k s_{k-1}^p |e_{k-1}|^{p^2}}{s_{k-1} |e_{k-1}|^p |e_{k-1}|} = s_k s_{k-1}^{p-1} |e_{k-1}|^{p^2-p-1}.$$

Since $p^2 - p - 1 = 0$, we have $|e_{k-1}|^{p^2-p-1} = 1$ and

$$|r_k| = s_k s_{k-1}^{p-1}.$$

Let $\rho_k = \log |r_k|$ and $\sigma_k = \log s_k$. Then our problem is to show that the sequence defined by

$$\sigma_k = \rho_k - (p-1)\sigma_k$$

has a limit.

Let $\rho_* = \lim_{k \to \infty} \rho_k$. Then the limit σ_*, if it exists, must satisfy

$$\sigma_* = \rho_* - (p-1)\sigma_*.$$

Thus we must show that the sequence of errors defined by

$$(\sigma_k - \sigma_*) = (\rho_k - \rho_*) - (p-1)(\sigma_k - \sigma_*)$$

converges to zero.

15. The convergence of the errors to zero can easily be established from first principles. However, with an eye to generalizations I prefer to use the following result from the theory of difference equations.

> If the roots of the equation
>
> $$x^n - a_1 x^{n-1} - \cdots - a_n = 0$$
>
> all lie in the unit circle and $\lim_{k \to \infty} \eta_k = 0$, then the sequence $\{\epsilon_k\}$ generated by the recursion
>
> $$\epsilon_k = \eta_k + a_1 \epsilon_{k-1} + \cdots a_n \epsilon_{k-n}$$
>
> converges to zero, whatever the starting values $\epsilon_0, \ldots, \epsilon_{n-1}$.

16. In our application $n = 1$ and $\epsilon_k = \sigma_k - \sigma_*$, and $\eta_k = \rho_k - \rho_*$. The equation whose roots are to lie in the unit circle is $x + (p-1) = 0$. Since $p - 1 \cong 0.618$, the conditions of the above result are satisfied, and $\sigma_k \to \sigma_*$. It follows that the numbers s_k have a nonzero limit. In other words, two-point convergence is superlinear of order $p = 1.618\ldots$.

Multipoint methods

17. The theory we developed for the secant method generalizes to multipoint iterations of the form

$$x_{k+1} = \varphi(x_k, x_{k-1}, \ldots, x_{k-n+1}).$$

Again the basic assumption is that if one of the arguments is the answer x_* then the value of φ is x_*. Under this assumption we can show that if the starting points are near enough x_* then the errors satisfy

$$\lim_{k \to \infty} \frac{e_{k+1}}{e_k e_{k-1} \cdots e_{k-n+1}} = \varphi_{12\ldots n}(x_*, x_*, \ldots, x_*),$$

where the subscript i of φ denotes differentiation with respect to the ith argument.

18. If $\varphi_{12\ldots n}(x_*, x_*, \ldots, x_*) \neq 0$, we say that the sequence exhibits *n-point convergence*. As we did earlier, we can show that n-point convergence is the same as pth-order convergence, where p is the largest root of the equation

$$p^n - p^{n-1} - \cdots - p - 1.$$

The following is a table of the convergence rates as a function of n.

n	p
2	1.61
3	1.84
4	1.93
5	1.96

The upper bound on the order of convergence is two, which is effectively attained for $n = 3$. For this reason multipoint methods of order four or greater are seldom encountered.

Muller's method

19. The secant method is sometimes called an interpolatory method, because it approximates a zero of a function by a line interpolating the function at two points. A useful iteration, called Muller's method, can be obtained by fitting a quadratic polynomial at three points. In outline, the iteration proceeds as follows. The input is three points x_k, x_{k-1}, x_{k-2}, and the corresponding function values.

1. Find a quadratic polynomial $g(x)$ such that $g(x_i) = f(x_i)$, ($i = k, k-1, k-2$).
2. Let x_{k+1} be the zero of g that lies nearest x_k.

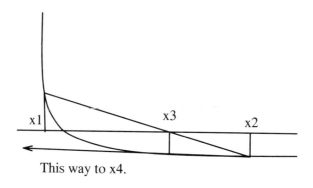

Figure 4.2. *A horrible example.*

It is a worthwhile exercise to work out the details.

20. Muller's method has the advantage that it can produce complex iterates from real starting values. This feature is not shared by either Newton's method or the secant method.

The linear-fractional method

21. Functions like the one pictured in Figure 4.2 come up occasionally, and they are difficult to solve. The figure shows the course of the secant method starting from a bracket $[x_1, x_2]$. The third iterate x_3 joins x_2 to the right of the zero, and because the function is flat there, x_4 is large and negative.

22. The trouble with the secant method in this case is that a straight line is not a good approximation to a function that has a vertical asymptote, followed by a zero and then a horizontal asymptote. On the other hand, the function

$$g(x) = \frac{x-a}{bx-c}$$

has a vertical asymptote at $x = \frac{c}{b}$, a zero at $x = a$, and a horizontal asymptote at $y = b^{-1}$ and therefore should provide a better approximation.

Since there are three free parameters in the function g, it is determined by three points. Thus a three-point interpolatory method similar to Muller's method can be based on linear-fractional functions.

23. In implementing the linear-fractional method it is easy to get lost in the details and end up with an indecipherable mess. My first encounter with the method was as a single FORTRAN statement that extended over several cards (Yes, cards!), and I was awed by the ingenuity of the programmer. It was only some years later that I learned better. By deriving the results we need in simple steps, we can effectively write a little program as we go along. Here is how it is done.

4. Nonlinear Equations

24. Most low-order interpolation problems are simplified by shifting the origin. In particular we take $y_i = x_i - x_k$ ($i = k, k-1, k-2$) and determine a, b, and c so that
$$g(y) = \frac{y-a}{by-c}$$
satisfies
$$f(x_i) = g(y_i), \quad i = k, k-1, k-2,$$
or equivalently
$$y_i - a = f(x_i)(by_i - c), \quad i = k, k-1, k-2. \tag{4.11}$$
Then $y_k = 0$, and the next point is given by
$$x_{k+1} = x_k + a.$$

25. Since at any one time there are only three points, there is no need to keep the index k around. Thus we start with three points x0, x1, x2, and their corresponding function values f0, f1, f2. We begin by setting
$$\begin{aligned} \text{y0} &= \text{x0} - \text{x2}, \\ \text{y1} &= \text{x1} - \text{x2}. \end{aligned}$$
From (4.11) we have
$$\begin{aligned} \text{y0} - \text{a} &= \text{f0}(\text{b} * \text{y0} - \text{c}), \\ \text{y1} - \text{a} &= \text{f1}(\text{b} * \text{y1} - \text{c}), \end{aligned}$$
and
$$\text{a} = \text{f2} * \text{c}. \tag{4.12}$$
If we add this last equation to the proceeding two we get
$$\begin{aligned} \text{y0} &= \text{fy0} * \text{b} + \text{df0} * \text{c}, \\ \text{y1} &= \text{fy1} * \text{b} + \text{df1} * \text{c}, \end{aligned} \tag{4.13}$$
where
$$\begin{aligned} \text{fy0} &= \text{f0} * \text{y0}, \\ \text{fy1} &= \text{f1} * \text{y1}, \end{aligned}$$
and
$$\begin{aligned} \text{df0} &= \text{f2} - \text{f0}, \\ \text{df1} &= \text{f2} - \text{f1}. \end{aligned}$$
The equations (4.13) can be solved for c by Cramer's rule:
$$\text{c} = \frac{\text{fy0} * \text{y1} - \text{fy1} * \text{y0}}{\text{fy0} * \text{df1} - \text{fy1} * \text{df0}}.$$

Thus from (4.12) the next iterate is

$$x3 = x2 + f2 * c.$$

26. Because we have chosen our origin carefully and have taken care to define appropriate intermediate variables, the above development leads directly to the following simple program. The input is the three points x0, x1, x2, and their corresponding function values f0, f1, f2. The output is the next iterate x3.

```
y0 = x0 - x2;
y1 = x1 - x2;
fy0 = f0*y0;
fy1 = f1*y1;
df0 = f2 - f0;
df1 = f2 - f1;
c = (fy0*y1-fy1*y0)/(fy0*df1-fy1*df0);
x3 = x2 + f2*c;
```

Lecture 5

Nonlinear Equations

A Hybrid Method
Errors, Accuracy, and Condition Numbers

A hybrid method

1. The secant method has the advantage that it converges swiftly and requires only one function evaluation per iteration. It has the disadvantage that it can blow up in your face. This can happen when the function is very flat so that $f'(x)$ is small compared with $f(x)$ (see Figure 4.2). Newton's method is also susceptible to this kind of failure; however, the secant method can fail in another way that is uniquely its own.

2. The problem is that in practice the function f will be evaluated with error. Specifically, the program that evaluates f at the point x will return not $f(x)$ but $\tilde{f}(x) = f(x) + e(x)$, where $e(x)$ is an unknown error. As long as $f(x)$ is large compared to $e(x)$, this error will have little effect on the course of the iteration. However, as the iteration approaches x_*, $e(x)$ may become larger than $f(x)$. Then the approximation to f' that is used in the secant method will have the value

$$\frac{[f(x_k) - f(x_{k-1})] + [e(x_k) - e(x_{k-1})]}{x_k - x_{k-1}}.$$

Since the terms in e dominate those in f, the value of this approximate derivative will be unpredictable. It may have the wrong sign, in which case the secant method may move away from x_*. It may be very small compared to $\tilde{f}(x_k)$, in which case the iteration will take a wild jump. Thus, if the function is computed with error, the secant method may behave erratically in the neighborhood of the zero it is supposed to find.

3. We are now going to describe a wonderful combination of the secant method and interval bisection.[5] The idea is very simple. At any stage of the iteration we work with three points a, b, and c. The points a and b are the points from which the next secant approximation will be computed; that is, they correspond to the points x_k and x_{k-1}. The points b and c form a proper bracket for the zero. If the secant method produces an undesirable approximation, we take the midpoint of the bracket as our next iterate. In

---

[5] The following presentation owes much to Jim Wilkinson's elegant technical report "Two Algorithms Based on Successive Linear Interpolation," Computer Science, Stanford University, TR CS-60, 1967.

this way the speed of the secant method is combined with the security of the interval bisection method. We will now fill in the details.

4. Let fa, fb, and fc denote the values of the function at a, b, and c. These function values are required to satisfy

$$
\begin{aligned}
&1.\ \ \mathtt{fa, fb, fc} \neq 0, \\
&2.\ \ \mathrm{sign}(\mathtt{fb}) \neq \mathrm{sign}(\mathtt{fc}), \\
&3.\ \ |\mathtt{fb}| \leq |\mathtt{fc}|.
\end{aligned}
\quad (5.1)
$$

At the beginning of the algorithm the user will be required to furnish points b and c = a satisfying the first two of these conditions. The user must also provide a convergence criterion eps. When the algorithm is finished, the bracketing points b and c will satisfy $|c - b| \leq$ eps.

5. The iterations take place in an endless while loop, which the program leaves upon convergence. Although the user must see that the first two conditions in (5.1) are satisfied, the program can take care of the third condition, since it has to anyway for subsequent iterations. In particular, if $|\mathtt{fc}| < |\mathtt{fb}|$, we interchange b and c. In this case, a and b may no longer be a pair of successive secant iterates, and therefore we set a equal to c.

```
while(1){
   if (abs(fc) < abs(fb))
   {
      t = c; c = b; b = t;
      t = fc; fc = fb; fb = t;                (5.2)
      a = c; fa = fc;
   }
```

6. We now test for convergence, leaving the loop if the convergence criterion is met.

```
   if (abs(b-c) <= eps)
      break;
```

7. The first step of the iteration is to compute the secant step s at the points a and b and also the midpoint m of b and c. One of these is to become our next iterate. Since $|\mathtt{fb}| \leq |\mathtt{fc}|$, it is natural to expect that x_* will be nearer to b than c, and of course it should lie in the bracket. Thus if s lies between b and m, then the next iterate will be s; otherwise it will be m.

8. Computing the next iterate is a matter of some delicacy, since we cannot say a priori whether b is to the left or right of c. It is easiest to cast the tests in terms of the differences ds = s − b and m = m − b. The following code does the trick. When it is finished, dd has been cómputed so that the next iterate is b + dd. Note the test to prevent division by zero in the secant step.

5. Nonlinear Equations

```
    dm = (c-b)/2;
    df =  (fa-fb);
    if (df == 0)
       ds = dm;
    else
       ds = -fb*(a-b)/df;
    if (sign(ds)!=sign(dm) || abs(ds) > abs(dm))
       dd = dm;
    else
       dd = ds;
```

9. At this point we make a further adjustment to dd. The explanation is best left for later (§5.16).

```
    if (abs(dd) < eps)
       dd = 0.5*sign(dm)*eps;
```

10. The next step is to form the new iterate—call it d—and evaluate the function there.

```
    d  = b + dd;
    fd = f(d);
```

11. We must now rename our variables in such a way that the conditions of (5.1) are satisfied. We take care of the condition that fd be nonzero by returning if it is zero.

```
    if (fd == 0){
       b = c = d; fb = fc = fd;
       break;
    }
```

12. Before taking care of the second condition in (5.1), we make a provisional assignment of new values to a, b, and c.

```
    a = b; b = d;
    fa = fb; fb = fd;
```

13. The second condition in (5.1) says that b and c form a bracket for x_*. If the new values fail to do so, the cure is to replace c by the *old* value of b. The reasoning is as follows. The old value of b has a different sign than the old value of c. The new value of b has the same sign as the old value of c. Consequently, the replacement results in a new value of c that has a different sign than the new value of b.

In making the substitution, it is important to remember that the old value of b is now contained in a.

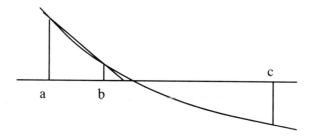

Figure 5.1. *A problem with* c.

```
if (sign(fb) == sign(fc)){
    c = a; fc = fa;
}
```

14. The third condition in (5.1) is handled at the top of the loop; see (5.2).

15. Finally, we return after leaving the `while` loop.

```
    }
    return;
```

16. To explain the adjustment of dd in §5.9, consider the graph in Figure 5.1. Here d is always on the side of x_* that is opposite c, and the value of c is not changed by the iteration. This means that although b is converging superlinearly to x_*, the length of the bracket converges to a number that is greater than zero — presumably much greater than eps. Thus the algorithm cannot converge until its erratic asymptotic behavior forces some bisection steps.

The cure for this problem lies in the extra code introduced in §5.9. If the step size dd is less than eps in absolute value, it is forced to have magnitude 0.5*eps. This will usually be sufficient to push s across the zero to the same side as c, which insures that the next bracket will be of length less than eps — just what is needed to meet the convergence criterion.

Errors, accuracy, and condition numbers

17. We have already observed in §5.2 that when we attempt to evaluate the function f at a point x the value will not be exact. Instead we will get a perturbed value

$$\tilde{f}(x) = f(x) + e(x).$$

The error $e(x)$ can come from many sources. It may be due to rounding error in the evaluation of the function, in which case it will behave irregularly. On the other hand, it may be dominated by approximations made in the evaluation of

5. Nonlinear Equations

the function. For example, an integral in the definition of the function may have been evaluated numerically. Such errors are often quite smooth. But whether or not the error is irregular or smooth, it is unknown and has an effect on the zeros of f that cannot be predicted. However, if we know something about the size of the error, we can say something about how accurately we can determine a particular zero.

18. Let x_* be a zero of f, and suppose we have a bound ϵ on the size of the error; i.e.,
$$|e(x)| \leq \epsilon.$$
If x_1 is a point for which $f(x_1) > \epsilon$, then
$$\tilde{f}(x_1) = f(x_1) + e(x_1) \geq f(x) - \epsilon > 0;$$
i.e., $\tilde{f}(x_1)$ has the same sign as $f(x_1)$. Similarly, if $f(x_2) < -\epsilon$, then $\tilde{f}(x_2)$ is negative along with $f(x_2)$, and by the intermediate value theorem f has a zero between x_1 and x_2. Thus, whenever $|f(x)| > \epsilon$, the values of $\tilde{f}(x)$ say something about the location of the zero in spite of the error.

To put the point another way, let $[a, b]$ be the largest interval about x_* for which
$$x \in [a, b] \implies f(x) \leq \epsilon.$$
As long as we are outside that interval, the value of $\tilde{f}(x)$ provides useful information about the location of the zero. However, inside the interval $[a, b]$ the value of $\tilde{f}(x)$ tells us nothing, since it could be positive, negative, or zero, regardless of the sign of $f(x)$.

19. The interval $[a, b]$ is an interval of uncertainty for the zero x_*: we know that x_* is in it, but there is no point in trying to pin it down further. Thus, a good algorithm will return a point in $[a, b]$, but we should not expect it to provide any further accuracy. Algorithms that have this property are called *stable* algorithms.

20. The size of the interval of uncertainty varies from problem to problem. If the interval is small, we say that the problem is *well conditioned*. Thus, a stable algorithm will solve a well-conditioned problem accurately. If the interval is large, the problem is *ill conditioned*. No algorithm, stable or otherwise, can be expected to return an accurate solution to an ill-conditioned problem. Only if we are willing to go to extra effort, like reducing the error $e(x)$, can we obtain a more accurate solution.

21. A number that quantifies the degree of ill-conditioning of a problem is called a *condition number*. To derive a condition number for our problem, let us compute the half-width of the interval of uncertainty $[a, b]$ under the assumption that
$$f'(x_*) \neq 0.$$

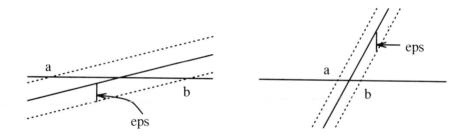

Figure 5.2. *Ill- and well-conditioned roots.*

From the approximation

$$f(x) \cong f(x_*) + f'(x_*)(x - x_*) = f'(x_*)(x - x_*)$$

it follows that $|f(x)| \leq \epsilon$ when $|f'(x_*)(x - x_*)| \lesssim \epsilon$. Hence,

$$|x - x_*| \lesssim \frac{\epsilon}{|f'(x_*)|},$$

or equivalently

$$[a, b] \cong \left[x_* - \frac{\epsilon}{|f'(x_*)|}, x_* + \frac{\epsilon}{|f'(x_*)|} \right].$$

Thus the number $1/|f'(x_*)|$ tells us how much the error is magnified in the solution and serves as a condition number. A zero with a large derivative is well conditioned: one with a small derivative is ill conditioned. Figure 5.2 illustrates these facts.

FLOATING-POINT ARITHMETIC

Lecture 6

Floating-Point Arithmetic

Floating-Point Numbers
Overflow and Underflow
Rounding Error
Floating-Point Arithmetic

Floating-point numbers

1. Anyone who has worked with a scientific hand calculator is familiar with floating-point numbers. Right now the display of my calculator contains the characters
$$2.597 \ -03 \tag{6.1}$$
which represent the number
$$2.597 \cdot 10^{-3}.$$

The chief advantage of floating-point representation is that it can encompass numbers of vastly differing magnitudes. For example, if we confine ourselves to six digits with five after the decimal point, then the largest number we can represent is $9.99999 \cong 10$, and the smallest is $0.00001 \cong 10^{-5}$. On the other hand, if we allocate two of those six digits to represent a power of ten, then we can represent numbers ranging between 10^{-99} and 10^{99}. The price to be paid is that these floating-point numbers have only four figures of accuracy, as opposed to as much as six for the fixed-point numbers.

2. A base-β floating-point number consists of a *fraction* f containing the significant figures of the number and *exponent* e containing its scale.[6] The value of the number is
$$f \cdot \beta^e.$$

3. A floating-point number $a = f \cdot \beta^e$ is said to be *normalized* if
$$\beta^{-1} \leq f < 1.$$

In other words, a is normalized if the base-β representation of its fraction has the form
$$f = 0.x_1 x_2 \ldots,$$
where $x_1 \neq 0$. Most computers work chiefly with normalized numbers, though you may encounter unnormalized numbers in special circumstances.

The term "normalized" must be taken in context. For example, by our definition the number (6.1) from my calculator is not normalized, while the

---
[6]The fraction is also called the *mantissa* and the exponent the *characteristic*.

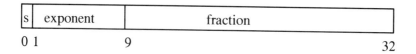

Figure 6.1. *A floating-point word.*

number $0.2597 \cdot 10^{-2}$ is. This does not mean that there is something wrong with my calculator—just that my calculator, like most, uses a different normalization in which $1 \leq f < 10$.

4. Three bases for floating-point numbers are in common use.

name	base	where found
binary	2	most computers
decimal	10	most hand calculators
hex	16	IBM mainframes and clones

In most computers, binary is the preferred base because, among other things, it fully uses the bits of the fraction. For example, the binary representation of the fraction of the hexadecimal number one is .00010000.... Thus, this representation wastes the three leading bits to store quantities that are *known* to be zero.

5. Even binary floating-point systems differ, something that in the past has made it difficult to produce portable mathematical software. Fortunately, the IEEE has proposed a widely accepted standard, which most PCs and workstations use. Unfortunately, some manufacturers with an investment in their own floating-point systems have not switched. No doubt they will eventually come around, especially since the people who produce mathematical software are increasingly reluctant to jury-rig their programs to conform to inferior systems.

6. Figure 6.1 shows the binary representation of a 32-bit IEEE standard floating-point word. One bit is devoted to the sign of the fraction, eight bits to the exponent, and twenty-three bits to the fraction. This format can represent numbers ranging in size from roughly 10^{-38} to 10^{38}. Its precision is about seven significant decimal digits. A curiosity of the system is that the leading bit of a normalized number is not represented, since it is known to be one.

The shortest floating-point word in a system is usually called a *single precision number*. Double precision numbers are twice as long. The double precision IEEE standard devotes one bit to the sign of the fraction, eleven bits to the exponent, and fifty-two bits to the fraction. This format can represent numbers ranging from roughly 10^{-307} to 10^{307} to about fifteen significant figures. Some implementations provide a 128-bit floating-point word, called a *quadruple precision number*, or *quad* for short.

Overflow and underflow

7. Since the set of real numbers is infinite, they cannot all be represented by a word of finite length. For floating-point numbers, this limitation comes in two flavors. First, the range of the exponent is limited; second, the fraction can only represent a finite number of the numbers between β^{-1} and one. The first of these limitations leads to the phenomena of overflow and underflow, collectively known as *exponent exceptions*; the second leads to rounding error. We will begin with exponent exceptions.

8. Whenever an arithmetic operation produces a number with an exponent that is too large, the result is said to have *overflowed*. For example, in a decimal floating-point system with a two-digit exponent, the attempt to square 10^{60} will result in overflow.

Similarly an arithmetic operation that produces an exponent that is too small is said to have *underflowed*. The attempt to square 10^{-60} will result in underflow in a decimal floating-point system with a two-digit exponent.

9. Overflow is usually a fatal error which will cause many systems to stop with an error message. In IEEE arithmetic, the overflow will produce a special word standing for infinity, and if the execution is continued it will propagate. An underflow is usually set to zero and the program continues to run. The reason is that with proper scaling, overflows can often be eliminated at the cost of generating harmless underflows.

10. To illustrate this point, consider the problem of computing

$$c = \sqrt{a^2 + b^2},$$

where $a = 10^{60}$ and $b = 1$. On a four-digit decimal computer the correctly rounded answer is 10^{60}. However, if the exponent has only two digits, then the computation will overflow while computing a^2.

The cure is to compute c in the form

$$c = s\sqrt{\left(\frac{a}{s}\right)^2 + \left(\frac{b}{s}\right)^2},$$

where s is a suitable scaling factor, say

$$s = \max\{|a|, |b|\} = 10^{60}.$$

With this value of s, we must compute

$$c = 10^{60}\sqrt{1^2 + \left(\frac{1}{10^{60}}\right)^2}.$$

Now when $(1/10^{60})^2$ is squared, it underflows and is set to zero. This does no harm, because 10^{-120} is insignificant compared with the number one, to which it is added. Continuing the computation, we obtain $c = 10^{60}$, which is what we wanted.

Although the above example is a simple calculation, it illustrates a technique of wide applicability — one that should be used more often.

Rounding error

11. The second limitation that a fixed word length imposes on floating-point arithmetic is that most numbers cannot be represented exactly. For example, the square root of seven is

$$2.6457513\ldots.$$

On a five-digit decimal computer, the digits beyond the fourth must be discarded. There are two traditional ways of doing this: conventional *rounding*, which yields

$$2.6458,$$

and *chopping* (also called *truncation*), which yields

$$2.6457.$$

12. It is instructive to compute a bound on the relative error that rounding introduces. The process is sufficiently well illustrated by rounding to five digits. Thus consider the number

$$a = \text{X.XXXXY},$$

which is rounded to

$$b = \text{X.XXXZ}.$$

Let us say we round up if $Y \geq 5$ and round down if $Y < 5$. Then it is easy to see that

$$|b - a| \leq 5 \cdot 10^{-5}.$$

On the other hand, the leading digit of a is assumed nonzero, and hence $|a| \geq 1$. It follows that

$$\frac{|b-a|}{|a|} \leq 5 \cdot 10^{-5} = \frac{1}{2} 10^{-4}.$$

More generally, rounding a to t decimal digits gives a number b satisfying

$$\frac{|b-a|}{|a|} = \frac{1}{2} 10^{-t+1}.$$

13. The same argument can be used to show that when a is chopped it gives a number b satisfying

$$\frac{|b-a|}{|a|} = 10^{-t+1}.$$

This bound is twice as large as the bound for rounding, as might be expected. However, as we shall see later, there are other, more compelling reasons for preferring rounding to chopping.

14. The bounds for t-digit binary numbers are similar:

$$\frac{|b-a|}{|a|} = \begin{cases} 2^{-t} & \text{rounding,} \\ 2^{-t+1} & \text{chopping.} \end{cases}$$

15. These bounds can be put in a form that is more useful for rounding-error analysis. Let $b = \text{fl}(a)$ denote the result of rounding or chopping a on a particular machine, and let ϵ_M denote the upper bound on the relative error. If we set

$$\epsilon = \frac{b-a}{a},$$

then $b = a(1+\epsilon)$ and $|\epsilon| \leq \epsilon_M$. In other words,

$$\text{fl}(a) = a(1+\epsilon), \qquad |\epsilon| \leq \epsilon_M. \tag{6.2}$$

16. The number ϵ_M in (6.2) is characteristic of the floating-point arithmetic of the machine in question and is called the *rounding unit* for the machine (also called *machine epsilon*). In some computations we may need to know the rounding unit. Many programming languages have library routines that return machine characteristics, such as the rounding unit or the exponent range. However, it is also possible to compute a reasonable approximation.

The key observation is that the rounding unit is at most slightly greater than the largest number x for which the computed value of 1 + x is equal to one. For example, if in six-digit binary arithmetic with rounding we take x = 2^{-7}, then the rounded value of 1 + x is exactly one. Thus, we can obtain an approximation to ϵ_M by starting with a large value of x and diminishing it until 1 + x evaluates to one. The following fragment does just that.

```
x = 1;
while (1+x != 1)
   x = x/2;
```

Unfortunately, this code can be undone by well-meaning but unsophisticated compilers that optimize the test 1+x != 1 to x != 0 or perform the test in extended precision registers.

Floating-point arithmetic

17. Most computers provide instructions to add, subtract, multiply, and divide floating-point numbers, and those that do not usually have software to do the same thing. In general a combination of floating-point numbers will not be

representable as a floating-point number of the same size. For example, the product of two five-digit numbers will generally require ten digits for its representation. Thus, the result of a floating-point operation can be represented only approximately.

18. Ideally, the result of a floating-point operation should be the exact result correctly rounded. More precisely, if $\mathrm{fl}(a \circ b)$ denotes the result of computing $a \circ b$ in floating-point and ϵ_M is the rounding unit, then we would like to have (cf. §6.15)
$$\mathrm{fl}(a \circ b) = (a \circ b)(1 + \epsilon), \qquad |\epsilon| \leq \epsilon_M.$$
Provided no exponent exceptions occur, the IEEE standard arithmetic satisfies this bound. So do most other floating-point systems, at least when $\circ = \times, \div$. However, some systems can return a difference with a large relative error, and it is instructive to see how this can come about.

19. Consider the computation of the difference $1 - 0.999999$ in six-digit decimal arithmetic. The first step is to align the operands thus:

$$\begin{array}{r} 1.000000 \\ -0.999999 \\ \hline \end{array}$$

If the computation is done in *seven*-digit registers, the computer can go on to calculate

$$\begin{array}{r} 1.000000 \\ -0.999999 \\ \hline 0.000001 \end{array}$$

and normalize the result to the correct answer: $.100000 \cdot 10^{-6}$. However, if the computer has only *six*-digit registers, the trailing 9 will be lost during the alignment. The resulting computation will proceed as follows

$$\begin{array}{r} 1.00000 \\ -0.99999 \\ \hline 0.00001 \end{array} \qquad (6.3)$$

giving a normalized answer of $.100000 \cdot 10^{-5}$. In this case, the computed answer has a relative error of ten!

20. The high relative error in the difference is due to the absence of an extra *guard digit* in the computation. Unfortunately some computer manufacturers fail to include a guard digit, and their floating-point systems are an annoyance to people who have to hold to high standards, e.g., designers of library routines for special functions. However, the vast majority of people never notice the absence of a guard digit, and it is instructive to ask why.

21. The erroneous answer $.100000 \cdot 10^{-5}$ computed in (6.3) is the difference between 1 and 0.99999 instead of 1 and 0.999999. Now the relative error in

6. Floating-Point Arithmetic

0.99999 as an approximation to 0.999999 is about $9 \cdot 10^{-6}$, which is of the same order of magnitude as the rounding unit ϵ_M. This means that the computed result could have been obtained by first making a very slight perturbation in the arguments and then performing the subtraction exactly. We can express this mathematically by saying that

$$\text{fl}(a \pm b) = a(1 + \epsilon_a) \pm b(1 + \epsilon_b), \qquad |\epsilon_a|, |\epsilon_b| \leq \epsilon_M. \qquad (6.4)$$

(Note that we may have to adjust the rounding unit upward a little for this bound to hold.)

22. Equation (6.4) is an example of a backward error analysis. Instead of trying to predict the error in the computation, we have shown that whatever the error it could have come from very slight perturbations in the original data. The power of the method lies in the fact that data usually comes equipped with errors of its own, errors that are far larger than the rounding unit. When put beside these errors, the little perturbations in (6.4) are insignificant. If the computation gives an unsatisfactory result, it is due to ill-conditioning in the problem itself, not to the computation.

We will return to this point later. But first an example of a nontrivial computation.

Lecture 7

Floating-Point Arithmetic

Computing Sums
Backward Error Analysis
Perturbation Analysis
Cheap and Chippy Chopping

Computing sums

1. The equation $\text{fl}(a+b) = (a+b)(1+\epsilon)$ ($|\epsilon| \leq \epsilon_M$) is the simplest example of a rounding-error analysis, and its simplest generalization is to analyze the computation of the sum

$$s_n = \text{fl}(x_1 + x_2 + \cdots + x_n).$$

There is a slight ambiguity in this problem, since we have not specified the order of summation. For definiteness assume that the x's are summed left to right.

2. The tedious part of the analysis is the repeated application of the error bounds. Let

$$s_i = \text{fl}(x_1 + x_2 + \cdots + x_i).$$

Then

$$s_2 = \text{fl}(x_1 + x_2) = (x_1 + x_2)(1 + \epsilon_1) = x_1(1 + \epsilon_1) + x_2(1 + \epsilon_1),$$

where $|\epsilon_1| \leq \epsilon_M$. Similarly,

$$\begin{aligned} s_3 = \text{fl}(s_2 + x_3) &= (s_2 + x_3)(1 + \epsilon_2) \\ &= x_1(1 + \epsilon_1)(1 + \epsilon_2) + \\ &\quad x_2(1 + \epsilon_1)(1 + \epsilon_2) + \\ &\quad x_3(1 + \epsilon_2). \end{aligned}$$

Continuing in this way, we find that

$$\begin{aligned} s_n = \text{fl}(s_{n-1} + x_n) &= (s_{n-1} + x_n)(1 + \epsilon_{n-1}) \\ &= x_1(1 + \epsilon_1)(1 + \epsilon_2) \cdots (1 + \epsilon_{n-1}) + \\ &\quad x_2(1 + \epsilon_1)(1 + \epsilon_2) \cdots (1 + \epsilon_{n-1}) + \\ &\quad x_3(1 + \epsilon_2) \cdots (1 + \epsilon_{n-1}) + \\ &\quad \vdots \\ &\quad x_{n-1}(1 + \epsilon_{n-2})(1 + \epsilon_{n-1}) + \\ &\quad x_n(1 + \epsilon_{n-1}), \end{aligned}$$

(7.1)

where $|\epsilon_i| \leq \epsilon_M$ ($i = 1, 2, \ldots, n-1$).

3. The expression (7.1) is not very informative, and it will help to introduce some notation. Let the quantities η_i be defined by

$$\begin{aligned}
1+\eta_1 &= (1+\epsilon_1)(1+\epsilon_2)\cdots(1+\epsilon_{n-1}), \\
1+\eta_2 &= (1+\epsilon_1)(1+\epsilon_2)\cdots(1+\epsilon_{n-1}), \\
1+\eta_3 &= (1+\epsilon_2)\cdots(1+\epsilon_{n-1}), \\
&\quad\vdots \\
1+\eta_{n-1} &= (1+\epsilon_{n-2})(1+\epsilon_{n-1}), \\
1+\eta_n &= (1+\epsilon_{n-1}).
\end{aligned}$$

Then

$$s_n = x_1(1+\eta_1)+x_2(1+\eta_2)+x_3(1+\eta_3)+\cdots+x_{n-1}(1+\eta_{n-1})+x_n(1+\eta_n). \quad (7.2)$$

4. The number $1+\eta_i$ is the product of numbers $1+\epsilon_j$ that are very near one. Thus we should expect that $1+\eta_i$ is itself near one. To get an idea of how near, consider the product

$$1+\eta_{n-1} = (1+\epsilon_{n-2})(1+\epsilon_{n-1}) = 1+(\epsilon_{n-2}+\epsilon_{n-1})+\epsilon_{n-2}\epsilon_{n-1}. \quad (7.3)$$

Now $|\epsilon_{n-2}+\epsilon_{n-1}| \leq 2\epsilon_M$ and $|\epsilon_{n-2}\epsilon_{n-1}| \leq \epsilon_M^2$. If, say, $\epsilon_M = 10^{-15}$, then $2\epsilon_M = 2\cdot 10^{-15}$ while $\epsilon_M^2 = 10^{-30}$. Thus the third term on the right-hand side of (7.3) is insignificant compared to the second term and can be ignored. If we ignore it, we get

$$\eta_{n-1} \cong \epsilon_{n-2}+\epsilon_{n-1}$$

or

$$|\eta_{n-1}| \lesssim |\epsilon_{n-2}|+|\epsilon_{n-1}| \leq 2\epsilon_M.$$

In general,

$$\begin{aligned}
|\eta_1| &\lesssim (n-1)\epsilon_M, \\
|\eta_i| &\lesssim (n-i+1)\epsilon_M, \qquad i=2,3,\ldots,n.
\end{aligned} \quad (7.4)$$

5. The approximate bounds (7.4) are good enough for government work, but there are fastidious individuals who *will* insist on rigorous inequalities. For them we quote the following result.

> If $n\epsilon_M \leq 0.1$ and $\epsilon_i \leq \epsilon_M$ ($i=1,2,\ldots,n$), then
> $$(1+\epsilon_1)(1+\epsilon_2)\cdots(1+\epsilon_n) = 1+\eta,$$
> where
> $$\eta \leq 1.06 n\epsilon_M.$$

7. Floating-Point Arithmetic

Thus if we set
$$\epsilon'_M = 1.06\epsilon_M,$$
then the approximate bounds (7.4) become quite rigorously
$$|\eta_1| \leq (n-1)\epsilon'_M,$$
$$|\eta_i| \leq (n-i+1)\epsilon'_M, \qquad i = 2, 3, \ldots, n. \qquad (7.5)$$

The quantity ϵ'_M is sometimes called the *adjusted rounding unit*.

6. The requirement that $n\epsilon_M \leq 0.1$ is a restriction on the size of n, and it is reasonable to ask if it is one we need to worry about. To get some idea of what it means, suppose that $\epsilon_M = 10^{-15}$. Then for this inequality to fail we must have $n \geq 10^{14}$. If we start summing numbers on a computer that can add at the rate of $1\mu\text{sec} = 10^{-6}\text{sec}$, then the time required to sum 10^{14} numbers is
$$10^8 \text{ sec} = 3.2 \text{ years}.$$

In other words, don't hold your breath waiting for $n\epsilon_M$ to become greater than 0.1.

Backward error analysis

7. The expression
$$s_n = x_1(1+\eta_1) + x_2(1+\eta_2) + x_3(1+\eta_3) + \cdots + x_{n-1}(1+\eta_{n-1}) + x_n(1+\eta_n), \quad (7.6)$$
along with the bounds on the η_i, is called a backward error analysis because the rounding errors made in the course of the computation are projected backward onto the original data. An algorithm that has such an analysis is called *stable* (or sometimes *backward stable*).

We have already mentioned in connection with the sum of two numbers (§6.22) that stability in the backward sense is a powerful property. Usually the backward errors will be very small compared to errors that are already in the input. In that case it is the latter errors that are responsible for the inaccuracies in the answer, not the rounding errors introduced by the algorithm.

8. To emphasize this point, suppose you are a numerical analyst and are approached by a certain Dr. XYZ who has been adding up some numbers.

> XYZ: I've been trying to compute the sum of ten numbers, and the answers I get are nonsense, at least from a scientific viewpoint. I wonder if the computer is fouling me up.
>
> YOU: Well it certainly has happened before. What precision were you using?
>
> XYZ: Double. I understand that it is about fifteen decimal digits.
>
> YOU: Quite right. Tell me, how accurately do you know the numbers you were summing?

XYZ: Pretty well, considering that they are experimental data. About four digits.

YOU: Then it's not the computer that is causing your poor results.

XYZ: How can you say that without even looking at the numbers? Some sort of magic?

YOU: Not at all. But first let me ask another question.

XYZ: Shoot.

YOU: Suppose I took your numbers and twiddled them in the sixth place. Could you tell the difference?

XYZ: Of course not. I already told you that we only know them to four places.

YOU: Then what would you say if I told you that the errors made by the computer could be accounted for by twiddling your data in the fourteenth place and then performing the computations exactly?

XYZ: Well, I find it hard to believe. But supposing it's true, you're right. It's my data that's the problem, not the computer.

9. At this point you might be tempted to bow out. Don't. Dr. XYZ wants to know more.

XYZ: But what went wrong? Why are my results meaningless?

YOU: Tell me, how big are your numbers?

XYZ: Oh, about a million.

YOU: And what is the size of your answer?

XYZ: About one.

YOU: And the answers you compute are at least an order of magnitude too large.

XYZ: How did you know that. Are you a mind reader?

YOU: Common sense, really. You have to cancel five digits to get your answer. Now if you knew your numbers to six or more places, you would get one or more accurate digits in your answer. Since you know only four digits, the lower two digits are garbage and won't cancel. You'll get a number in the tens or greater instead of a number near one.

XYZ: What you say makes sense. But does that mean I have to remeasure my numbers to six or more figures to get what I want?

YOU: That's about it.

XYZ: Well I suppose I should thank you. But under the circumstances, it's not easy.

YOU: That's OK. It comes with the territory.

10. The above dialogue is artificial in three respects. The problem is too simple to be characteristic of real life, and no scientist would be as naive as

7. Floating-Point Arithmetic

Dr. XYZ. Moreover, people don't role over and play dead like Dr. XYZ: they require a lot of convincing. But the dialogue illustrates two important points.

The first point is that a backward error analysis is a useful tool for removing the computer as a suspect when something goes wrong. The second is that backward stability is seldom enough. We want to know *what* went wrong. What is there in the problem that is causing difficulties? To use the terminology we introduced for zeros of functions: When is the problem ill-conditioned?

Perturbation analysis

11. To answer the question just posed, it is a good idea to drop any considerations of rounding error and ask in general what effects known errors in the x_i will have on the sum

$$\sigma = x_1 + x_2 + \cdots + x_n.$$

Specifically, we will suppose that

$$\tilde{x}_i = x_i(1 + \mu_i), \qquad |\mu_i| \leq \epsilon, \tag{7.7}$$

and look for a bound on the error in the sum

$$\tilde{\sigma} = \tilde{x}_1 + \tilde{x}_2 + \cdots + \tilde{x}_n.$$

Such a procedure is called a *perturbation analysis* because it assesses the effects of perturbations in the arguments of a function on the value of the function.

The analysis is easy enough to do. We have

$$|\tilde{\sigma} - \sigma| \leq |x_1||\mu_1| + |x_2||\mu_2| + \cdots + |x_n||\mu_n|.$$

From (7.7), we obtain the following bound on the absolute error:

$$|\tilde{\sigma} - \sigma| \leq (|x_1| + |x_2| + \cdots + |x_n|)\epsilon.$$

We can now obtain a bound on the relative error by dividing by $|\sigma|$. Specifically, if we set

$$\kappa = \frac{|x_1| + |x_2| + \cdots + |x_n|}{|x_1 + x_2 + \cdots + x_n|},$$

then

$$\frac{|\tilde{\sigma} - \sigma|}{|\sigma|} \leq \kappa\epsilon. \tag{7.8}$$

The number κ, which is never less than one, tells how the rounding errors made in the course of the computation are magnified in the result. Thus it serves as a condition number for the problem. (Take a moment to look back at the discussion of condition numbers in §5.21.)

12. In Dr. XYZ's problem, the experimental errors in the fourth place can be represented by $\epsilon = 10^{-4}$, in which case the bound becomes

$$\text{relative error} = \kappa \cdot 10^{-4}.$$

Since there were ten x's of size about 5,000,000, while the sum of the $x'a$ was about one, we have $\kappa \cong \cdot 10^7$, and the bound says that we can expect no accuracy in the result, regardless of any additional rounding errors.

13. We can also apply the perturbation analysis to bound the effects of rounding errors on the sum. In this case the errors μ_i correspond to the errors η_i in (7.2). Thus from (7.5) we have

$$|\mu_i| \leq (n-1)\epsilon'_M \equiv \epsilon.$$

It then follows from (7.8) that

$$\frac{|s_n - \sigma|}{|\sigma|} = (n-1)\kappa\epsilon'_M,$$

where as usual

$$\kappa = \frac{|x_1| + |x_2| + \cdots + |x_n|}{|x_1 + x_2 + \cdots + x_n|}$$

is the condition number for the sum.

Cheap and chippy chopping

14. When the terms x_i in the sum are all positive (or all negative), the condition number κ is one; i.e., the problem is perfectly conditioned. In this case, the bound on the relative error due to rounding reduces to

$$\frac{|s_n - \sigma|}{|\sigma|} = (n-1)\epsilon'_M.$$

This inequality predicts that rounding error will accumulate slowly as terms are added to the sum. However, the analysis on which the bound was based assumes that the worst happens all the time, and one might expect that the factor $n-1$ is an overestimate.

In fact, if we sum positive numbers with rounded arithmetic, the factor *will* be an overestimate, since the individual rounding errors will be positive or negative at random and will tend to cancel one other. On the other hand, if we are summing positive numbers with chopped arithmetic, the errors will tend to be in the same direction (downward), and they will reinforce one another. In this case the factor $n-1$ is realistic.

15. We don't have to resort to a lengthy analysis to see how this phenomenon comes about. Instead, let's imagine that we take two six-digit numbers, and

7. Floating-Point Arithmetic

do two things with them. First, we round the numbers to five digits and sum them exactly; second, we chop the numbers to five digits and once again sum them exactly. The following table shows what happens.

	number	=	rounded	+	error	=	chopped	+	error
	1374.8	=	1375	−	0.2	=	1374	+	0.8
	3856.4	=	3856	+	0.4	=	3856	+	0.4
total	5231.2	=	5231	+	0.2	=	5230	+	1.2

As can be seen from the table, the errors made in rounding have opposite signs and cancel each other in the sum to yield a small error of 0.2. With chopping, however, the errors have the same sign and reinforce each other to yield a larger error of 1.2. Although we have summed only two numbers to keep things simple, the errors in sums with more terms tend to behave in the same way: errors from rounding tend to cancel, while errors from chopping reinforce. Thus rounding is to be preferred in an algorithm in which it may be necessary to sum numbers all having the same sign.

16. The above example makes it clear that you cannot learn everything about a floating-point system by studying the bounds for its arithmetic. In the bounds, the difference between rounding and chopping is a simple factor of two, yet when it comes to sums of positive numbers the difference in the two arithmetics is a matter of the accumulation of errors. In particular, the factor $n-1$ in the error bound (7.8) reflects how the error may grow for chopped arithmetic, while it is unrealistic for rounded arithmetic. (On statistical grounds it can be argued that the factor $\sqrt{n}$ is realistic for rounded arithmetic.)

To put things in a different light, binary, chopped arithmetic has the same bound as binary, rounded arithmetic with one less bit. Yet on the basis of what we have seen, we would be glad to sacrifice the bit to get the rounding.

Lecture 8

Floating-Point Arithmetic

Cancellation
The Quadratic Equation
That Fatal Bit of Rounding Error
Envoi

Cancellation

1. Many calculations seem to go well until they have to form the difference between two nearly equal numbers. For example, if we attempt to calculate the sum

$$37654 + 25.874 - 37679 = 0.874$$

in five-digit floating-point, we get

$$\text{fl}(37654 + 25.874) = 37680$$

and

$$\text{fl}(37860 - 37679) = 1.$$

This result does not agree with the true sum to even one significant figure.

2. The usual explanation of what went wrong is to say that we cancelled most of the significant figures in the calculation of fl(37860 − 37679) and therefore the result cannot be expected to be accurate. Now this is true as far as it goes, but it conveys the mistaken impression that the cancellation *caused* the inaccuracy. However, if you look closely, you will see that no error at all was made in calculating fl(37860 − 37679). Thus the source of the problem must lie elsewhere, and the cancellation simply revealed that the computation was in trouble.

In fact, the source of the trouble is in the addition that preceded the cancellation. Here we computed fl(37654+25.874) = 37680. Now this computation is the same as if we had replaced 25.874 by 26 and computed 37654 + 26 exactly. In other words, this computation is equivalent to throwing out the three digits 0.874 in the number 25.874. Since the answer consists of just these three digits, it is no wonder that the final computed result is wildly inaccurate. What has killed us is not the cancellation but the loss of important information earlier in the computation. The cancellation itself is merely a death certificate.

The quadratic equation

3. To explore the matter further, let us consider the problem of solving the quadratic equation

$$x^2 - bx + c = 0,$$

whose roots are given by the quadratic formula

$$r = \frac{b \pm \sqrt{b^2 - 4c}}{2}.$$

If we take

$$b = 3.6778 \quad \text{and} \quad c = 0.0020798,$$

then the roots are

$$r_1 = 3.67723441190\ldots \quad \text{and} \quad r_2 = 0.00056558809\ldots.$$

4. An attempt to calculate the smallest root in five-digit arithmetic gives the following sequence of operations.

$$
\begin{array}{lll}
1. & b^2 & : 1.3526 \cdot 10^{+1} \\
2. & 4c & : 8.3192 \cdot 10^{-3} \\
3. & b^2 - 4c & : 1.3518 \cdot 10^{+1} \\
4. & \sqrt{b^2 - 4c} & : 3.6767 \cdot 10^{+0} \\
5. & b - \sqrt{b^2 - 4c} & : 1.1000 \cdot 10^{-3} \\
6. & (b - \sqrt{b^2 - 4c})/2 & : 5.5000 \cdot 10^{-4}
\end{array}
\tag{8.1}
$$

The computed value 0.00055000 differs from the true value 0.000565... of the root in its second significant figure.

5. According to the conventional wisdom on cancellation, the algorithm failed at step 5, where we canceled three-odd significant figures in computing the difference $3.6778 - 3.6667$. However, from the point of view taken here, the cancellation only reveals a loss of information that occurred earlier. In this case the offending operation is in step 3, where we compute the difference

$$\text{fl}(13.453 - 0.0083192) = 13.445.$$

This calculation corresponds to replacing the number 0.0083192 by 0.008 and performing the calculation exactly. This is in turn equivalent to replacing the coefficient $c = 0.0020798$ by $\tilde{c} = 0.002$ and performing the calculation exactly. Since the coefficient c contains critical information about r_2, it is no wonder that the change causes the computed value of r_2 to be inaccurate.

6. Can anything be done to save the algorithm? It depends. If we don't save c after using it in step 3, then the answer is we can do nothing: the numbers that we have at hand, namely b and the computed value of $b^2 - 4c$, simply do not have the information necessary to recover an accurate value of r_2. On the other hand, if we keep c around, then we can do something.

8. Floating-Point Arithmetic

7. The first thing to observe is that there is no problem in calculating the largest root r_1, since taking the plus sign in the quadratic formula entails no cancellation. Thus after step 4 in (8.1) we can proceed as follows.

$$\begin{array}{lll} 5.' & b + \sqrt{b^2 - 4c} & : \quad 7.3545 \cdot 10^{+0} \\ 6.' & r_1 = (b + \sqrt{b^2 - 4c})/2 & : \quad 3.6773 \cdot 10^{+0} \end{array}$$

The result agrees with the true value of r_1 to almost the last place.

To calculate r_2, we next observe that $c = r_1 r_2$, so that

$$r_2 = \frac{c}{r_1}.$$

Since we have already computed r_1 accurately, we may use this formula to recover r_2 as follows.

$$7. \quad r_2 = c/r_1 \quad : \quad 5.6558 \cdot 10^{-4}$$

The computed value is as accurate as we can reasonably expect.

8. Many calculations in which cancellation occurs can be salvaged by rewriting the formulas. The major exceptions are intrinsically ill-conditioned problems that are poorly defined by their data. The problem is that the inaccuracies that were built into the problem will have to reveal themselves one way or another, so that any attempt to suppress cancellation at one point will likely introduce it at another.

For example, the discriminant $b^2 - 4c$ is equal to $(r_1 - r_2)^2$, so that cancellation in its computation is an indication that the quadratic has nearly equal roots. Since $f'(r_1) = 2r_1 - b = r_1 - r_2$, these nearby roots will be ill conditioned (see §5.21).

That fatal bit of rounding error

9. Consider the behavior of the solution of the difference equation

$$x_{k+1} = 2.25 x_k - 0.5 x_{k-1}, \tag{8.2}$$

where

$$x_1 = \frac{1}{3} \quad \text{and} \quad x_2 = \frac{1}{12}. \tag{8.3}$$

The solution is

$$x_k = \frac{4^{1-k}}{3}, \quad k = 1, 2, 3, \ldots.$$

Consequently the computed solution should decrease indefinitely, each successive component being a fourth of its predecessor.

10. Figure 8.1 contains a graph of $\log_2 x_k$ as a function of k. Initially this

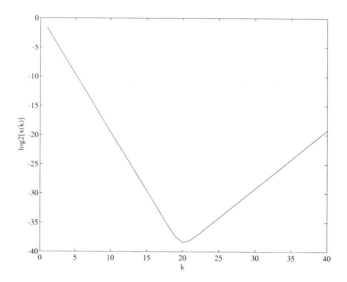

Figure 8.1. *Computed solution of* $x_{k+1} = 2.25x_k - 0.5x_{k-1}$.

graph descends linearly with a slope of -2, as one would expect of any function proportional to $(1/4)^k$. However, at $k = 20$ the graph turns around and begins to ascend with a slope of one. What has gone wrong?

11. The answer is that the difference equation (8.2) has two principal solutions:

$$\left(\frac{1}{4}\right)^k \quad \text{and} \quad 2^k.$$

Any solution can be expanded as a linear combination of these two solutions; i.e., the most general form of a solution is

$$\alpha\left(\frac{1}{4}\right)^k + \beta 2^k.$$

Now in principle, the x_k defined by (8.2) and (8.3) should have an expansion in which $\beta = 0$; however, because of rounding error, β is effectively nonzero, though very small. As time goes by, the influence of this solution grows until it dominates. Thus the descending part of the graph represents the interval in which the contribution of $\beta 2^k$ is negligible, while the ascending portion represents the interval in which $\beta 2^k$ dominates.

12. It is possible to give a formal rounding-error analysis of the computation of x_k. However, it would be tedious, and there is a better way of seeing

8. Floating-Point Arithmetic

what is going on. We simply assume that all the rounding error is made at the beginning of the calculation and that the remaining calculations are performed exactly.

Specifically, let us assume that errors made in rounding have given us x_1 and x_2 that satisfy

$$x_1 = \frac{1}{3}(4^{+0} + 2^{-56}),$$
$$x_2 = \frac{1}{3}(4^{-1} + 2^{-55})$$

(note that 2^{-56} is the rounding unit for IEEE 64-bit arithmetic). Then the general solution is

$$x_k = \frac{1}{3}(4^{1-k} + 2^{k-57}).$$

The turnaround point for this solution occurs when

$$4^{1-k} = 2^{k-57},$$

which gives a value of k between nineteen and twenty. Obviously, our simplified analysis has predicted the results we actually observed.

13. All this illustrates a general technique of wide applicability. It frequently happens that an algorithm has a critical point at which a little bit of rounding error will cause it to fail later. If you think you know the point, you can confirm it by rounding at that point but allowing no further rounding errors. If the algorithm goes bad, you have spotted a weak point, since it is unlikely that the rounding errors you have *not* made will somehow correct your fatal bit of error.

Envoi

14. We have now seen three ways in which rounding error can manifest itself.

1. Rounding error can accumulate, as it does during the computation of a sum. Such accumulation is slow and is usually important only for very long calculations.

2. Rounding error can be revealed by cancellation. The occurrence of cancellation is invariably an indication that something went wrong earlier in the calculation. Sometimes the problem can be cured by changing the details of the algorithm; however, if the source of the cancellation is an intrinsic ill-conditioning in the problem, then it's back to the drawing board.

3. Rounding error can be magnified by an algorithm until it dominates the numbers we actually want to compute. Again the calculation does not have to be lengthy. There are no easy fixes for this kind of problem.

It would be wrong to say that these are the only ways in which rounding error makes itself felt, but they account for many of the problems observed in practice. If you think you have been bitten by rounding error, you could do worse that ask if the problem is one of the three listed above.

LINEAR EQUATIONS

Lecture 9

Linear Equations

Matrices, Vectors, and Scalars
Operations with Matrices
Rank-One Matrices
Partitioned Matrices

Matrices, vectors, and scalars

1. An $m \times n$ *matrix* A is a rectangular array of numbers of the form

$$A = \begin{pmatrix} a_{11} & a_{12} & \cdots & a_{1,n-1} & a_{1n} \\ a_{21} & a_{22} & \cdots & a_{2,n-1} & a_{2n} \\ \vdots & \vdots & & \vdots & \vdots \\ a_{n-1,1} & a_{n-1,2} & \cdots & a_{n-1,n-1} & a_{n-1,n} \\ a_{n1} & a_{n2} & \cdots & a_{n,n-1} & a_{nn} \end{pmatrix}.$$

We write $A \in \mathbf{R}^{m \times n}$. If $m = n$, so that A is square, we say that A is of *order* n.

2. The numbers a_{ij} are called the *elements* of the matrix A. By convention, the first subscript, i, called the *row index*, indicates the row in which the element lies. The second subscript, j, called the *column index*, indicates the column in which the element lies. Indexing usually begins at one, which makes for minor problems in the language C, whose arrays begin with an index of zero.

3. An *n-vector* x is an array of the form

$$x = \begin{pmatrix} x_1 \\ x_2 \\ \vdots \\ x_n \end{pmatrix}.$$

We write $x \in \mathbf{R}^n$. The number n is called the dimension. The numbers x_j are called the *components* of x.

4. Note that by convention, all vectors are column vectors; that is, their components are arranged in a column. Objects like $(x_1 \; x_2 \; \cdots \; x_n)$ whose components are arranged in a row are called *row vectors*. We generally write row vectors in the form x^{T} (see the definition of the transpose operation below in §9.18).

5. We will make no distinctions between $\mathbf{R}^{n \times 1}$ and $\mathbf{R}^n$: it is all the same to us whether we call an object an $n \times 1$ matrix or an n-vector. Similarly, we will

not distinguish between the real numbers **R**, also called *scalars*, and the set of 1-vectors, and the set of 1×1 matrices.

6. Matrices will be designated by upper-case Latin or Greek letters, e.g., A, Λ, etc. Vectors will be designated by lower-case Latin letters, e.g., x, y, etc. Scalars will be designated by lower-case Latin and Greek letters. Some attempt will be made to use an associated lower-case letter for the elements of a matrix or the components of a vector. Thus the elements of A will be a_{ij} or possibly α_{ij}. In particular note the association of ξ with x and η with y.

Operations with matrices

7. Matrices are more than static arrays of numbers: they have an algebra. We will be particularly concerned with the following four operations with matrices:

 1. multiplication by a scalar,
 2. the matrix sum,
 3. the matrix product,
 4. the matrix transpose.

8. Any matrix A can be multiplied by a scalar μ. The result is the matrix μA defined by
$$\mu A = (\mu a_{ij}).$$

9. If A and B have the same dimensions, then their sum is the matrix $A + B$ defined by
$$A + B = (a_{ij} + b_{ij}).$$
We express the fact that A and B have the same dimensions by saying that their dimensions are *conformal* for summation.

10. A matrix whose elements are all zero is called a zero matrix and is written 0 regardless of its dimensions. It is easy to verify that
$$A + 0 = 0 + A = 0,$$
so that 0 is an additive identity for matrices.

11. If A is an $l \times m$ matrix and B is an $m \times n$ matrix, then the product AB is an $l \times n$ matrix defined by
$$AB = \left(\sum_{k=1}^{m} a_{ik} b_{kj} \right).$$

9. Linear Equations

Note that for the product AB to be defined, the number of columns of A must be the same as the number of rows of B. In this case we say that the dimensions conform for multiplication.

12. The matrix product is so widely used that it is useful to have a recipe for doing the bookkeeping. Here is mine. To find the $(2,3)$-element of the matrix product

$$\begin{pmatrix} a_{11} & a_{12} & a_{13} & a_{14} \\ a_{21}^1 & a_{22}^2 & a_{23}^3 & a_{24}^4 \\ a_{31} & a_{32} & a_{33} & a_{34} \end{pmatrix} \begin{pmatrix} b_{11} & b_{12} & b_{13}^1 & b_{14} & b_{15} \\ b_{21} & b_{22} & b_{23}^2 & b_{24} & b_{25} \\ b_{31} & b_{32} & b_{33}^3 & b_{34} & b_{35} \\ b_{41} & b_{42} & b_{43}^4 & b_{44} & b_{45} \end{pmatrix},$$

place your left index finger on a_{21} and your right index finger on b_{13} (these are the elements with the superscript one). As you do so, say, "Times." Now move your fingers to a_{22} and b_{23}, saying "plus" as you move them and "times" as you land. Continue in this manner, alternating "plus" and "times." At the end you will have computed

$$a_{21}b_{13} + a_{22}b_{23} + a_{23}b_{33} + a_{24}b_{43},$$

which is the $(2,3)$-element. You may feel foolish doing this, but you'll get the right answer.

13. The matrix I_n of order n whose diagonal elements are one and whose off-diagonal elements are zero [we write $I_n = \text{diag}(1, 1, \ldots, 1)$] is called the *identity matrix*. If A is any $m \times n$ matrix, then it is easy to verify that

$$I_m A = A I_n,$$

so that identity matrices are multiplicative identities. When the context makes the order clear, we drop the subscript and simply write I for an identity matrix.

14. The identity matrix is a special case of a useful class of matrices called *diagonal matrices*. A matrix D is diagonal if its only nonzero entries lie on its diagonal, i.e., if $d_{ij} = 0$ whenever $i \neq j$. We write

$$\text{diag}(d_1, \ldots, d_n)$$

for a diagonal matrix whose diagonal entries are $d_1, \ldots, d_n$.

15. Since we have agreed to regard an n-vector as an $n \times 1$ matrix, the above definitions can be transferred directly to vectors. Any vector can be multiplied by a scalar. Two vectors of the same dimension may be added. Only 1-vectors, i.e., scalars, can be multiplied.

16. A particularly important case of the matrix product is the matrix-vector product Ax. Among other things it is useful as an abbreviated way of writing

systems of equations. Rather than say that we shall solve the system

$$b_1 = a_{11}x_1 + a_{12}x_2 + \cdots + a_{1n}x_n$$
$$b_2 = a_{21}x_1 + a_{22}x_2 + \cdots + a_{2n}x_n$$
$$\cdot \quad \cdot \quad \cdot$$
$$b_n = a_{n1}x_1 + a_{n2}x_2 + \cdots + a_{nn}x_n$$

we can simply write that we shall solve the equation

$$b = Ax,$$

where A is of order n.

17. Both the matrix sum and the matrix product are associative; that is, $(A + B) + C = A + (B + C)$ and $(AB)C = A(BC)$. The product distributes over the sum; e.g., $A(B + C) = AB + AC$. In addition, the matrix sum is commutative: $A + B = B + A$. Unfortunately the matrix product is not commutative: in general $AB \neq BA$. It is easy to forget this fact when you are manipulating formulas involving matrices.

18. The final operation we shall use is the matrix transpose. If A is an $m \times n$ matrix, then the transpose of A is the $n \times m$ matrix A^T defined by

$$A^T = (a_{ji}).$$

Thus the transpose is the matrix obtained by reflecting a matrix through its diagonal.

19. The transpose interacts nicely with the other matrix operations:

1. $(\mu A)^T = \mu(A^T)$,
2. $(A + B)^T = A^T + B^T$,
3. $(AB)^T = B^T A^T$.

Note that the transposition reverses the order of a product.

20. If x is a vector, then x^T is a row vector. If x and y are n-vectors, then

$$y^T x = x_1 y_1 + x_2 y_2 + \cdots + x_n y_n$$

is a scalar called the *inner product* of x and y. In particular the number

$$\|x\| = \sqrt{x^T x}$$

is the Euclidean length (or two-norm) of the vector x.

Rank-one matrices

21. If $x, y \neq 0$, then any matrix of the form

$$W = xy^{\mathrm{T}} = \begin{pmatrix} x_1 y_1 & x_1 y_2 & x_1 y_3 & \cdots \\ x_2 y_1 & x_2 y_2 & x_2 y_3 & \cdots \\ x_3 y_1 & x_3 y_2 & x_3 y_3 & \cdots \\ \vdots & \vdots & \vdots & \end{pmatrix} \qquad (9.1)$$

has rank one; that is, its columns span a one-dimensional space. Conversely, any rank-one matrix W can be represented in the form xy^{T}. Rank-one matrices arise frequently in numerical applications, and it's important to know how to deal with them.

22. The first thing to note is that one does not store a rank-one matrix as a matrix. For example, if x and y are n-vectors, then the matrix xy^{T} requires n^2 locations to store, as opposed to $2n$ locations to store x and y. To get some idea of the difference, suppose that $n = 1000$. Then xy^{T} requires one million words to store as a matrix, as opposed to 2000 to store x and y individually — the storage differs by a factor of 500.

23. If we always represent a rank-one matrix $W = xy^{\mathrm{T}}$ by storing x and y, the question arises of how we perform matrix operations with W — how, say, we can compute the matrix-vector product $c = Wb$? An elegant answer to this question may be obtained from the equation

$$c = Wb = (xy^{\mathrm{T}})b = x(y^{\mathrm{T}}b) = (y^{\mathrm{T}}b)x, \qquad (9.2)$$

in which the last equality follows from the fact that $y^{\mathrm{T}}b$ is a scalar.

This equation leads to the following algorithm.

1. Compute $\mu = y^{\mathrm{T}}b$
2. Compute $c = \mu x$ $\qquad (9.3)$

This algorithm requires $2n$ multiplications and $n-1$ additions. This should be contrasted with the roughly n^2 multiplications and additions required to form an ordinary matrix vector product.

24. The above example illustrates the power of matrix methods in deriving efficient algorithms. A person contemplating the full matrix representation (9.1) of xy^{T} would no doubt come up with what amounts to the algorithm (9.3), albeit in scalar form. But the process would be arduous and error prone. On the other hand, the simple manipulations in (9.2) yield the algorithm directly and in a way that relates it naturally to operations with vectors. We shall see further examples of the power of matrix techniques in deriving algorithms.

Partitioned matrices

25. Partitioning is a device by which we can express matrix operations at a level between scalar operations and operations on full matrices. A *partition* of a matrix is a decomposition of the matrix into submatrices. For example, consider the matrix

$$A = \begin{pmatrix} a_{11} & a_{12} & a_{13} & a_{14} & a_{15} & a_{16} & a_{17} \\ a_{21} & a_{22} & a_{23} & a_{24} & a_{25} & a_{26} & a_{27} \\ a_{31} & a_{32} & a_{33} & a_{34} & a_{35} & a_{36} & a_{37} \\ a_{41} & a_{42} & a_{43} & a_{44} & a_{45} & a_{46} & a_{47} \\ a_{51} & a_{52} & a_{53} & a_{54} & a_{55} & a_{56} & a_{57} \end{pmatrix}.$$

The partitioning induced by the lines in the matrix allows us to write the matrix in the form

$$A = \begin{pmatrix} A_{11} & A_{12} & A_{13} \\ A_{21} & A_{22} & A_{23} \end{pmatrix},$$

where

$$A_{11} = \begin{pmatrix} a_{11} & a_{12} \\ a_{21} & a_{22} \end{pmatrix}, \quad A_{12} = \begin{pmatrix} a_{13} & a_{14} & a_{15} \\ a_{23} & a_{24} & a_{25} \end{pmatrix}, \quad \text{etc.}$$

26. The power of partitioning lies in the following fact.

> If the partitions of two matrices are conformal, the submatrices may be treated as scalars for the purposes of performing matrix operations.

For example, if the partitions below are conformal, we have

$$\begin{pmatrix} A_{11} & A_{12} \\ A_{21} & A_{22} \end{pmatrix} + \begin{pmatrix} B_{11} & B_{12} \\ B_{21} & B_{22} \end{pmatrix} = \begin{pmatrix} A_{11}+B_{11} & A_{12}+B_{12} \\ A_{21}+B_{21} & A_{22}+B_{22} \end{pmatrix}.$$

Similarly, we may write a matrix product in the form

$$\begin{pmatrix} A_{11} & A_{12} \\ A_{21} & A_{22} \end{pmatrix} \begin{pmatrix} B_{11} & B_{12} \\ B_{21} & B_{22} \end{pmatrix} = \begin{pmatrix} A_{11}B_{11}+A_{12}B_{21} & A_{11}B_{12}+A_{12}B_{22} \\ A_{21}B_{11}+A_{22}B_{21} & A_{21}B_{12}+A_{22}B_{22} \end{pmatrix},$$

again provided the partitions are conformal. The one thing to be careful about here is that the products of the submatrices do not commute like scalar products.

27. As a simple but important example of the use of matrix partitions, consider the matrix-vector product Ax, where A is of order n. If we partition A by columns in the form

$$A = (a_1 \; a_2 \; \ldots \; a_n),$$

9. Linear Equations

then
$$Ax = (a_1 \ a_2 \ \ldots \ a_n) \begin{pmatrix} \xi_1 \\ \xi_2 \\ \vdots \\ \xi_n \end{pmatrix} = \xi_1 a_1 + \xi_2 a_2 + \cdots + \xi_n a_n.$$

From this formula we can draw the following useful conclusion.

> The matrix-vector product Ax is the linear combination of the columns of A whose coefficients are the components of x.

Lecture 10

Linear Equations

The Theory of Linear Systems
Computational Generalities
Triangular Systems
Operation Counts

The theory of linear systems

1. For some time to come we will be concerned with the solution of the equation

$$Ax = b, \qquad (10.1)$$

where A is of order n. Before attempting to solve *any* numerical problem it is a good idea to find out if it has a unique solution. Fortunately, the theory of linear systems provides a number of conditions that can be used to check if (10.1) has a solution.

> Let A be of order n. Then the following statements are equivalent.
>
> 1. For any vector b, the system $Ax = b$ has a solution.
> 2. If a solution of the system $Ax = b$ exists, it is unique.
> 3. For all x, $Ax = 0 \implies x = 0$.
> 4. The columns (rows) of A are linearly independent.
> 5. There is a matrix A^{-1} such that $A^{-1}A = AA^{-1} = I$.
> 6. $\det(A) \neq 0$.

2. Although the above conditions all have their applications, as a practical matter the condition $\det(A) \neq 0$ can be quite misleading. The reason is that the determinant changes violently with minor rescaling. Specifically, if A is of order n, then

$$\det(\sigma A) = \sigma^n \det(A).$$

To see the implications of this equality, suppose that $n = 30$ (rather small by today's standards) and that $\det(A) = 1$. Then

$$\det(0.1 \cdot A) = 10^{-30}.$$

In other words, dividing the elements of A by ten reduces the determinant by a factor of 10^{-30}. It is not easy to determine whether such a volatile quantity is truly zero—which is why the determinant is used primarily in theoretical settings.

3. A matrix A satisfying any of the conditions of §10.1 is said to be *nonsingular*. If A is nonsingular, the matrix A^{-1} guaranteed by the fifth condition is called the *inverse* of A. The inverse interacts nicely with matrix operations of multiplication and transposition.

Let A and B be of order n.

1. The product AB is nonsingular if and only if A and B are nonsingular. In this case
$$(AB)^{-1} = B^{-1}A^{-1}.$$

2. The matrix A^{T} is nonsingular and
$$(A^{\mathrm{T}})^{-1} = (A^{-1})^{\mathrm{T}}.$$

A convenient shorthand for $(A^{\mathrm{T}})^{-1}$ is $A^{-\mathrm{T}}$.

Computational generalities

4. The solution of the linear system $Ax = b$ can be written in terms of the matrix inverse as follows
$$A^{-1}b = A^{-1}(Ax) = Ix = x.$$

This suggests the following algorithm for solving linear systems.

1. Compute $C = A^{-1}$
2. $x = Cb$ (10.2)

With very few exceptions, this is a bad algorithm. In fact, you would not even use it for scalar equations. For example, if we were to try to compute the solution of $10x = 2$ in this way, we would end up with

1. $c = 1/10$
2. $x = 2c$

If instead we write $x = 2/10$, we save an operation and a little rounding error. The situation is the same with the general invert-and-multiply algorithm (10.2), except that (10.2) is much more expensive than its alternatives and a lot less stable.

5. What *are* the alternatives? Later we are going to show how to factor a matrix A in the form[7]
$$A = LU,$$

---

[7] This is a slight oversimplification. For stability we have to interchange rows of A as we factor it.

10. Linear Equations

where L is lower triangular (i.e., its elements are zero above the diagonal) and U is upper triangular (its elements are zero below the diagonal). This factorization is called an *LU decomposition* of the matrix A. Now if A is nonsingular, then so are L and U. Consequently, if we write the system $Ax = b$ in the form $LUx = b$, we have

$$Ux = L^{-1}b \equiv y. \tag{10.3}$$

Moreover, by the definition of y

$$Ly = b. \tag{10.4}$$

Thus, if we have a method for solving triangular systems, we can use the following algorithm to solve the system $Ax = b$.

1. Factor $A = LU$
2. Solve $Ly = b$
3. Solve $Ux = y$

To implement this general algorithm we must be able to factor A and to solve triangular systems. We will begin with triangular systems.

Triangular systems

6. A matrix L is *lower triangular* if

$$i < j \implies \ell_{ij} = 0.$$

This is a fancy way of saying that the elements of L lying above the diagonal are zero. For example, a lower triangular matrix of order five has the form

$$L = \begin{pmatrix} \ell_{11} & 0 & 0 & 0 & 0 \\ \ell_{21} & \ell_{22} & 0 & 0 & 0 \\ \ell_{31} & \ell_{32} & \ell_{33} & 0 & 0 \\ \ell_{41} & \ell_{42} & \ell_{43} & \ell_{44} & 0 \\ \ell_{51} & \ell_{52} & \ell_{53} & \ell_{54} & \ell_{55} \end{pmatrix}.$$

A matrix U is *upper triangular* if

$$i > j \implies u_{ij} = 0.$$

The elements of an upper triangular matrix lying below the diagonal are zero.

7. For definiteness we will consider the solution of the lower triangular system $Lx = b$ of order n. The basic fact about such systems is that if the diagonal elements ℓ_{ii} of L are nonzero, then L is nonsingular and the system has a unique solution. We shall establish this fact by describing an algorithm to compute the solution.

8. The algorithm is sufficiently well illustrated by the case $n = 5$. We begin by writing the system $Lx = b$ in scalar form.

$$\begin{aligned} b_1 &= \ell_{11}x_1 \\ b_2 &= \ell_{21}x_1 + \ell_{22}x_2 \\ b_3 &= \ell_{31}x_1 + \ell_{32}x_2 + \ell_{33}x_3 \\ b_4 &= \ell_{41}x_1 + \ell_{42}x_2 + \ell_{43}x_3 + \ell_{44}x_4 \\ b_5 &= \ell_{51}x_1 + \ell_{52}x_2 + \ell_{53}x_3 + \ell_{54}x_4 + \ell_{54}x_5 \end{aligned}$$

The first equation in this system involves only x_1 and may be solved forthwith:

$$x_1 = \frac{b_1}{\ell_{11}}.$$

Knowing the value of x_1, we can substitute it into the second equation and solve to get

$$x_2 = \frac{b_2 - \ell_{21}x_1}{\ell_{22}}.$$

Substituting x_1 and x_2 into the third equation and solving, we get

$$x_3 = \frac{b_3 - \ell_{31}x_1 - \ell_{32}x_2}{\ell_{33}}.$$

Continuing in this manner, we get

$$x_4 = \frac{b_4 - \ell_{41}x_1 - \ell_{42}x_2 - \ell_{43}x_3}{\ell_{44}}$$

and

$$x_5 = \frac{b_5 - \ell_{51}x_1 - \ell_{52}x_2 - \ell_{53}x_3 - \ell_{54}x_4}{\ell_{55}},$$

which completes the solution of the system. Since by hypothesis the diagonal elements ℓ_{ii} are nonzero, these formulas uniquely determine the x_i.

9. The procedure sketched above is quite general and leads to the following *forward-substitution algorithm* for solving a lower triangular system. The lower triangular matrix is contained in a doubly subscripted array l. The components of the solution overwrite the right-hand side b.[8]

```
for (i=1; i<=n; i++){
    for (j=1; j<i; j++)
        b[i] = b[i] - l[i][j]*b[j];
    b[i] = b[i]/l[i][i];
}
```
(10.5)

---

[8]We have already noted that the indexing conventions for matrices, in which the first element is the (1,1)-element, are inconsistent with C array conventions in which the first element of the array a is a[0][0]. In most C code presented here, we will follow the matrix convention. This wastes a little storage for the unused part of the array, but that is a small price to pay for consistency.

10. Linear Equations

Operation counts

10. To get an idea of how much it costs to solve a triangular system, let us count the number of multiplications required by the algorithm. There is one multiplication in the statement

```
b[i] = b[i] - l[i][j]*b[j];
```

This statement is executed for `j` running from 1 to `i` and for `i` running from 1 to `n`. Hence the total number of multiplications is

$$\sum_{i=1}^{n}\sum_{j=1}^{i} 1 = \sum_{i=1}^{n} i = \frac{n(n+1)}{2} \cong \frac{n^2}{2}, \qquad (10.6)$$

the last approximation holding for large n. There are a like number of additions.

11. Before we try to say what an operation count like (10.6) actually means, let us dispose of a technical point. In deriving operation counts for matrix processes, we generally end up with sums nested two or three deep, and we are interested in the dominant term, i.e., the term with the highest power of n. We can obtain this term by replacing sums with integrals and adjusting the limits of integration to make life easy. If this procedure is applied to (10.6), the result is

$$\int_0^n \int_0^i 1 \, dj \, di = \int_0^n i \, di = \frac{n^2}{2},$$

which is the dominant term in the sum (10.6).

12. It might be thought that one could predict the running time of an algorithm by counting its arithmetic operations and multiplying by the time required for an operation. For example, if the combined time for a floating-point addition and multiplication is α, then it might be expected that it would take time $\frac{n^2\alpha}{2}$ to solve a triangular system of order n.

Unfortunately, this expectation is not realized in practice. The reason is that the algorithm is busy with tasks other than floating-point arithmetic. For example, the reference to `l[i][j]` requires that an element of `l` be retrieved from memory. Again, the test `j<i` must be performed each time the inner loop is executed. This overhead inflates the time, so that an operation count based solely on arithmetic will underestimate the total time.

Nonetheless, operation counts of the kind we have introduced here can be useful. There are two reasons.

1. The additional overhead in an algorithm is generally proportional to the number of arithmetic operations. Although an arithmetic count does not predict the running time, it predicts how the running time will increase with n — linearly, quadratically, etc.

2. In consequence, if algorithms have different orders of complexity — that is, dominant terms with different powers of n — then the one with the lower order will ultimately run faster.

13. Some care must be taken in comparing algorithms of the same order. The presumption is that the one with the smaller order constant will run faster. Certainly, if the order constants differ significantly, this is a reasonable assumption: we would expect an algorithm with a count of $2n^2$ to outperform an algorithm with a count of $100n^2$. But when the order constants are nearly equal, all bets are off, since the actually order constant will vary according to the details of the algorithm. Someone who says that an algorithm with a count of n^2 is faster than an algorithm with a count of $2n^2$ is sticking out the old neck.

Lecture 11

Linear Equations

Memory Considerations
Row-Oriented Algorithms
A Column-Oriented Algorithm
General Observations
Basic Linear Algebra Subprograms

Memory considerations

1. Virtual memory is one of the more important advances in computer systems to come out of the 1960s. The idea is simple, although the implementation is complicated. The user is supplied with very large *virtual memory*. This memory is subdivided into blocks of modest size called *pages*. Since the entire virtual memory cannot be contained in fast, main memory, most of its pages are maintained on a slower backing store, usually a disk. Only a few active pages are contained in the main memory.

When an instruction references a memory location, there are two possibilities.

 1. The page containing the location is in main memory (a hit). In this case the location is accessed immediately.
 2. The page containing the location is not in main memory (a miss). In this case the system selects a page in main memory and swaps it with the one that is missing.

2. Since misses involve a time-consuming exchange of data between main memory and the backing store, they are to be avoided if at all possible. Now memory locations that are near one another are likely to lie on the same page. Hence one strategy for reducing misses is to arrange to access memory sequentially, one neighboring location after another. This is a special case of what is called *locality of reference*.

Row-oriented algorithms

3. The algorithm (10.5) is one that preserves locality of reference. The reason is that the language C stores doubly subscripted arrays by rows, so that in the case $n = 5$ the matrix l might be stored as follows.

$$\ell_{11}\, 0\, 0\, 0\, 0\, \ell_{21}\, \ell_{22}\, 0\, 0\, 0\, \ell_{31}\, \ell_{32}\, \ell_{33}\, 0\, 0\, \ell_{41}\, \ell_{42}\, \ell_{43}\, \ell_{44}\, 0\, \ell_{51}\, \ell_{52}\, \ell_{53}\, \ell_{54}\, \ell_{55}$$

Now if you run through the loops in (10.5), you will find that the elements of L are accessed in the following order.

$$\begin{array}{cccccc}
1 & 2\ 3 & 4\ 5\ 6 & 7\ 8\ 9\ 10 & 11\ 12\ 13\ 14\ 15 \\
\ell_{11}\ 0\ 0\ 0\ 0 & \ell_{21}\ \ell_{22}\ 0\ 0\ 0 & \ell_{31}\ \ell_{32}\ \ell_{33}\ 0\ 0 & \ell_{41}\ \ell_{42}\ \ell_{43}\ \ell_{44}\ 0 & \ell_{51}\ \ell_{52}\ \ell_{53}\ \ell_{54}\ \ell_{55}
\end{array}$$

Clearly the accesses here tend to be sequential. Once a row is in main memory, we march along it a word at a time.

4. A matrix algorithm like (10.5) in which the inner loops access the elements of the matrix by rows is said to be *row oriented*. Provided the matrix is stored by rows, as it is in C, row-oriented algorithms tend to interact nicely with virtual memories.

5. The situation is quite different with the language FORTRAN, in which matrices are stored by columns. For example, in FORTRAN the elements of the matrix L will appear in storage as follows.

$$\ell_{11}\ \ell_{21}\ \ell_{31}\ \ell_{41}\ \ell_{51}\ 0\ \ell_{22}\ \ell_{32}\ \ell_{42}\ \ell_{52}\ 0\ 0\ \ell_{33}\ \ell_{43}\ \ell_{53}\ 0\ 0\ 0\ \ell_{44}\ \ell_{54}\ 0\ 0\ 0\ 0\ \ell_{55}$$

If the FORTRAN equivalent of algorithm (10.5) is run on this array, the memory references will occur in the following order.

$$\begin{array}{ccccc}
1\ 2\ 4\ 7\ 11 & 3\ 5\ 8\ 12 & 6\ 9\ 13 & 10\ 14 & 15 \\
\ell_{11}\ \ell_{21}\ \ell_{31}\ \ell_{41}\ \ell_{51} & 0\ \ell_{22}\ \ell_{32}\ \ell_{42}\ \ell_{52} & 0\ 0\ \ell_{33}\ \ell_{43}\ \ell_{53} & 0\ 0\ 0\ \ell_{44}\ \ell_{54} & 0\ 0\ 0\ 0\ \ell_{55}
\end{array}$$

Clearly the references are jumping all over the place, and we can expect a high miss rate for this algorithm.

A column-oriented algorithm

6. The cure for the FORTRAN problem is to get another algorithm — one that is column oriented. Such an algorithm is easy to derive from a partitioned form of the problem.

Specifically, let the system $Lx = b$ be partitioned in the form

$$\begin{pmatrix} \lambda_{11} & 0 \\ \ell_{21} & L_{22} \end{pmatrix} \begin{pmatrix} \xi_1 \\ x_2 \end{pmatrix} \begin{pmatrix} \beta_1 \\ b_2 \end{pmatrix},$$

where L_{22} is lower triangular. (Note that we now use the Greek letter λ to denote individual elements of L, so that we do not confuse the vector $\ell_{21} = (\lambda_{21}, \ldots, \lambda_{n1})^{\mathrm{T}}$ with the element λ_{21}.) This partitioning is equivalent to the two equations

$$\begin{aligned} \lambda_{11}\xi_1 &= \beta_1, \\ \ell_{21}\xi_1 + L_{22}x_2 &= b_2. \end{aligned}$$

11. Linear Equations

The first equation can be solved as usual:

$$\xi_1 = \frac{\beta_1}{\lambda_{11}}.$$

Knowing the first component ξ_1 of x we can write the equation in the form

$$L_{22}x_2 = b_2 - \xi_1 \ell_{21}. \tag{11.1}$$

But this equation is a lower triangular system of order one less than the original. Consequently we can repeat the above procedure and reduce it to a lower triangular system of order two less than the original, and so on until we reach a 1×1 system, which can be readily solved.

7. The following FORTRAN code implements this algorithm.

$$\begin{array}{l}\texttt{do 20 j=1,n}\\ \quad\texttt{b(j) = b(j)/l(j,j)}\\ \quad\texttt{do 10 i=j+1,n}\\ \quad\quad\texttt{b(i) = b(i) - b(j)*l(i,j)}\\ \texttt{10}\quad\texttt{continue}\\ \texttt{20 continue}\end{array} \tag{11.2}$$

At the jth iteration of the outer loop, the components $x_1, \ldots, x_{j-1}$ have already been computed and stored in b(1), ..., b(j-1). Thus we have to solve the system involving the matrix

$$\begin{pmatrix} \lambda_{jj} & 0 \\ \ell_{j+1,j} & L_{j+1,j+1} \end{pmatrix}.$$

The statement

$$\texttt{b(j) = b(j)/l(j,j)}$$

overwrites b(j) with the first component of this solution. The loop

```
       do 10 i=j+1,n
          b(i) = b(i) + b(j)*l(i,j)
10     continue
```

then adjusts the right-hand side as in (11.1). The algorithm then continues to solve the smaller system whose matrix is $L_{j+1,j+1}$.

8. Running through the algorithm, we find that the memory references to the elements of L occur in the following order.

$$\begin{array}{cccccccccccccccc}1 & 2 & 3 & 4 & 5 & & 6 & 7 & 8 & 9 & & 10 & 11 & 12 & & 13 & 14 & & 15\\ \ell_{11} & \ell_{21} & \ell_{31} & \ell_{41} & \ell_{51} & 0 & \ell_{22} & \ell_{32} & \ell_{42} & \ell_{52} & 0\,0 & \ell_{33} & \ell_{43} & \ell_{53} & 0\,0\,0 & \ell_{44} & \ell_{54} & 0\,0\,0\,0 & \ell_{55}\end{array}$$

Clearly, the references here are much more localized than in the row-oriented algorithm, and we can expect fewer misses.

General observations on row and column orientation

9. Although we have considered row- and column-oriented algorithms in terms of their interactions with virtual memories, many computers have more than two levels of memory. For example, most computers have a limited amount of superfast memory called a *cache*. Small blocks of main memory are swapped in and out of the cache, much like the pages of a virtual memory. Obviously, an algorithm that has good locality of reference will use the cache more efficiently.

10. It is important to keep things in perspective. If the entire matrix can fit into main memory, the distinction between row- and column-oriented algorithms becomes largely a matter of cache effects, which may not be all that important. A megaword of memory can hold a matrix of order one thousand, and many workstations in common use today have even more memory.

To illustrate this point, I ran row- and column-oriented versions of Gaussian elimination (to be treated later in §13) coded in C on my SPARC IPC workstation. For a matrix of order 1000, the row-oriented version ran in about 690 seconds, while the column-oriented version took 805 seconds. Not much difference. On a DEC workstation, the times were 565 seconds for the row-oriented version and 1219 seconds for the column-oriented version. A more substantial difference, but by no means an order of magnitude.[9]

11. Finally, the problems treated here illustrate the need to know something about the target computer and its compilers when one is designing algorithms. An algorithm that runs efficiently on one computer may not run well on another. As we shall see, there are ways of hiding some of these machine dependencies in specially designed subprograms; but the general rule is that algorithms have to run on actual machines, and machines have idiosyncrasies which must be taken into account.

Basic linear algebra subprograms

12. One way of reducing the dependency of algorithms on the characteristics of particular machines is to perform frequently used operations by invoking subprograms instead of writing in-line code. The chief advantage of this approach is that manufacturers can provide sets of these subprograms that have been optimized for their individual machines. Subprograms to facilitate matrix computations are collectively known as Basic Linear Algebra Subprograms, or BLAS for short.

13. To illustrate the use of the BLAS, consider the loop

```
        for (j=1; j<i; j++)
            b[i] = b[i] - l[i][j]*b[j];
```

---

[9] The programs were compiled by the gcc compiler with the optimization flag set and were run on unloaded machines.

11. Linear Equations

from the algorithm (10.5). In expanded form, this computes
```
b[i] = b[i] - l[i][1]*b[1] - l[i][2]*b[2] - ...
            - l[i][i-1]*b[i-1];
```
that is, it subtracts the dot product of the vectors

$$\begin{pmatrix} l[i][1] \\ l[i][2] \\ \vdots \\ l[i][i-1] \end{pmatrix} \text{ and } \begin{pmatrix} b[1] \\ b[2] \\ \vdots \\ b[i-1] \end{pmatrix}$$

from b[i]. Consequently, if we write a little function
```
float dot(n, float x[], float y[])
{
   float d=0;
   for (i=0; i<n; i++)
      d = d + x[i]*y[i];
   return d;
}
```
we can rewrite (10.5) in the form
```
for (i=1; i<=n; i++)
   b[i] = (b[i] - dot(i-1, &l[i][1], &b[1]))/l[i][i];
```
Not only do we now have the possibility of optimizing the code for the function dot, but the row-oriented algorithm itself has been considerably simplified.

14. As another example, consider the following statements from the column-oriented algorithm (11.2).
```
      do 10 i=j+1,n
         b(i) = b(i) - b(j)*l(i,j)
10    continue
```
Clearly these statements compute the vector

$$\begin{pmatrix} b(j+1) \\ b(j+2) \\ \vdots \\ b(n) \end{pmatrix} - b(j) \begin{pmatrix} l(j+1,j) \\ l(j+2,j) \\ \vdots \\ l(n,j) \end{pmatrix}.$$

Consequently, if we write a little FORTRAN program axpy (for $ax + y$)
```
      subroutine axpy(n, a, x, y)
      integer n
      real a, x(*), y(*)
      do 10 i=1,n
         y(i) = y(i) + a*x(i)
10    continue
      return
      end
```

then we can rewrite the program in the form

```
      do 20 j=1,n
         b(j) = b(j)/l(j,j)
         call axpy(n-j, -b(j), l(j+1,j), b(j+1))
   20 continue
```

Once again the code is simplified, and we have the opportunity to optimize axpy for a given machine.

15. The programs dot and axpy are two of the most frequently used vector BLAS; however, there are many more. For example, scal multiplies a vector by a scalar, and copy copies a vector into the locations occupied by another. It should be stressed that the names and calling sequence we have given in these examples are deliberate oversimplifications, and you should read up on the actual BLAS before writing production code.

Lecture 12

Linear Equations

Positive-Definite Matrices
The Cholesky Decomposition
Economics

Positive-definite matrices

1. A matrix A of order n is *symmetric* if $A^\mathrm{T} = A$, or equivalently if

$$a_{ij} = a_{ji}, \quad i,j = 1,\ldots,n.$$

Because of this structure a symmetric matrix is entirely represented by the elements on and above its diagonal, and hence can be stored in half the memory required for a general matrix. Moreover, many matrix algorithms can be simplified for symmetric matrices so that they have smaller operation counts. Let us see what symmetry buys us for linear systems.

2. The best way to solve a linear system $Ax = b$ is to factor A in the form LU, where L is lower triangular and U is upper triangular and then solve the resulting triangular systems (see §10.5). When A is symmetric, it is reasonable to expect the factorization to be symmetric; that is, one should be able to factor A in the form $A = R^\mathrm{T} R$, where R is upper triangular. However, not just any symmetric matrix has such a factorization.

To see why, suppose A is nonsingular and $x \neq 0$. Then R is nonsingular, and $y = Rx \neq 0$. Hence,

$$x^\mathrm{T} A x = x^\mathrm{T} R^\mathrm{T} R x = (Rx)^\mathrm{T}(Rx) = y^\mathrm{T} y = \sum y_i^2 > 0. \qquad (12.1)$$

Thus a nonsingular matrix A that can be factored in the form $R^\mathrm{T} R$ has the following two properties.

$$\begin{array}{ll} 1. & A \text{ is symmetric.} \\ 2. & \cdot\ x \neq 0 \implies x^\mathrm{T} A x > 0. \end{array} \qquad (12.2)$$

Any matrix with these two properties is said to be *positive definite*.[10]

3. Positive-definite matrices occur frequently in real life. For example, a variant of the argument (12.1) shows that if X has linearly independent columns then $X^\mathrm{T} X$ is positive definite. This means, among other things, that positive-definite matrices play an important role in least squares and regression, where systems like $(X^\mathrm{T} X)b = c$ are common. Again, positive-definite matrices arise

---

[10]Warning: Some people drop symmetry when defining a positive-definite matrix.

in connection with elliptic partial differential equations, which occur everywhere. In fact, no advanced student in the hard sciences or engineering can fail to meet with a positive-definite matrix in the course of his or her studies.

4. Positive-definite matrices are nonsingular. To see this we will show that

$$Ax = 0 \implies x = 0$$

(see §10.1). Suppose on the contrary that $Ax = 0$ but $x \neq 0$. Then $0 = x^T A x$, which contradicts the positive-definiteness of A. Thus $x = 0$, and A is nonsingular.

5. Matrices lying on the diagonal of a partitioned positive-definite matrix are positive definite. In particular, if we partition A in the form

$$A = \begin{pmatrix} \alpha & a^T \\ a & A_* \end{pmatrix},$$

then $\alpha > 0$ and A_* is positive definite. To see that $\alpha > 0$, set $x = (1, 0, \ldots, 0)$. Then

$$0 < x^T A x = (1\ 0) \begin{pmatrix} \alpha & a^T \\ a & A_* \end{pmatrix} \begin{pmatrix} 1 \\ 0 \end{pmatrix} = \alpha.$$

To see that A_* is positive definite, let $y \neq 0$ and set $x^T = (0\ y^T)$. Then

$$0 < x^T A x = (0\ y^T) \begin{pmatrix} \alpha & a^T \\ a & A_* \end{pmatrix} \begin{pmatrix} 0 \\ y \end{pmatrix} = y^T A_* y.$$

The Cholesky decomposition

6. We have seen that any nonsingular matrix A that can be factored in the form $R^T R$ is positive definite. The converse is also true. If A is positive definite, then A can be factored in the form $A = R^T R$, where R is upper triangular. If, in addition, we require the diagonal elements of R to be positive, the decomposition is unique and is called the *Cholesky decomposition* or the *Cholesky factorization* of A. The matrix R is called the *Cholesky factor* of A. We are going to establish the existence and uniqueness of the Cholesky decomposition by giving an algorithm for computing it.

7. We will derive the Cholesky algorithm in the same way we derived the column-oriented method for solving lower triangular equations: by considering a suitable partition of the problem, we solve part of it and at the same time reduce it to a problem of smaller order. For the Cholesky algorithm the partition of the equation $A = R^T R$ is

$$\begin{pmatrix} \alpha & a^T \\ a & A_* \end{pmatrix} = \begin{pmatrix} \rho & 0 \\ r & R_*^T \end{pmatrix} \begin{pmatrix} \rho & r^T \\ 0 & R_* \end{pmatrix}.$$

12. Linear Equations

Writing out this equation by blocks of the partition, we get the three equations

1. $\alpha = \rho^2$,
2. $a^{\mathrm{T}} = \rho r^{\mathrm{T}}$,
3. $A_* = R_*^{\mathrm{T}} R_* + rr^{\mathrm{T}}$.

Equivalently,

1. $\rho = \sqrt{\alpha}$,
2. $r^{\mathrm{T}} = \rho^{-1} a^{\mathrm{T}}$, \hfill (12.3)
3. $R_*^{\mathrm{T}} R_* = A_* - rr^{\mathrm{T}}$.

The first two equations are, in effect, an algorithm for computing the first row of R. The (1,1)-element ρ of R is well defined, since $\alpha > 0$. Since $\rho \neq 0$, r^{T} is uniquely defined by the second equation.

The third equation says that R_* is the Cholesky factor of the matrix

$$\hat{A}_* = A_* - rr^{\mathrm{T}} = A_* - \alpha^{-1} a a^{\mathrm{T}}$$

[the last equality follows from the first two equations in (12.3)]. This matrix is of order one less than the original matrix A, and consequently we can compute its Cholesky factorization by applying our algorithm recursively. However, we must first establish that $\hat{A}_*$ is itself positive definite, so that it has a Cholesky factorization.

8. The matrix $\hat{A}_*$ is clearly symmetric, since

$$\hat{A}_*^{\mathrm{T}} = (A_* - rr^{\mathrm{T}})^{\mathrm{T}} = A_*^{\mathrm{T}} - (r^{\mathrm{T}})^{\mathrm{T}} r^{\mathrm{T}} = A_* - rr^{\mathrm{T}}.$$

Hence it remains to show that for any nonzero vector y

$$y^{\mathrm{T}} \hat{A} y = y^{\mathrm{T}} A_* y - \alpha^{-1} (a^{\mathrm{T}} y)^2 > 0.$$

To do this we will use the positive-definiteness of A. If η is any scalar, then

$$0 < (\eta\ y^{\mathrm{T}}) \begin{pmatrix} \alpha & a^{\mathrm{T}} \\ a & A_* \end{pmatrix} \begin{pmatrix} \eta \\ y \end{pmatrix} = \alpha \eta^2 + 2 a^{\mathrm{T}} y \eta + y^{\mathrm{T}} A_* y.$$

If we now set $\eta = \alpha^{-1} a^{\mathrm{T}} y$, then it follows after a little manipulation that

$$0 < \alpha \eta^2 + 2\eta a^{\mathrm{T}} y + y^{\mathrm{T}} A_* y = y^{\mathrm{T}} A_* y - \alpha^{-1} (a^{\mathrm{T}} y)^2,$$

which is what we had to show.

9. Before we code the algorithm sketched above, let us examine its relation to an elimination method for solving a system of equations $Ax = b$. We begin by writing the equation $\hat{A}_* = A_* - \alpha^{-1} a a^{\mathrm{T}}$ in scalar form as follows:

$$\hat{\alpha}_{ij} = \alpha_{ij} - \alpha_{11}^{-1} \alpha_{1i} \alpha_{1j} = \alpha_{ij} - \alpha_{11}^{-1} \alpha_{i1} \alpha_{1j}.$$

Here we have put the subscripting of A back into the partition, so that $\alpha = \alpha_{11}$ and $a^{\mathrm{T}} = (\alpha_{12}, \ldots, \alpha_{1n})$. The second equality follows from the symmetry of A.

Now consider the system

$$\alpha_{11}x_1 + \alpha_{12}x_2 + \alpha_{13}x_3 + \alpha_{14}x_4 = b_1$$
$$\alpha_{21}x_1 + \alpha_{22}x_2 + \alpha_{23}x_3 + \alpha_{24}x_4 = b_2$$
$$\alpha_{31}x_1 + \alpha_{32}x_2 + \alpha_{33}x_3 + \alpha_{34}x_4 = b_3$$
$$\alpha_{41}x_1 + \alpha_{42}x_2 + \alpha_{43}x_3 + \alpha_{44}x_4 = b_4$$

If the first equation is solved for x_1, the result is

$$x_1 = \alpha_{11}^{-1}(b_1 - \alpha_{12}x_2 - \alpha_{13}x_3 - \alpha_{14}x_4).$$

Substituting x_1 in the last three of the original equations and simplifying, we get

$$(\alpha_{22} - \alpha_{11}^{-1}\alpha_{21}\alpha_{12})x_2 + (\alpha_{23} - \alpha_{11}^{-1}\alpha_{21}\alpha_{13})x_3 + (\alpha_{24} - \alpha_{11}^{-1}\alpha_{21}\alpha_{14})x_4$$
$$= b_2 - \alpha_{11}^{-1}\alpha_{21}b_1$$
$$(\alpha_{32} - \alpha_{11}^{-1}\alpha_{31}\alpha_{12})x_2 + (\alpha_{33} - \alpha_{11}^{-1}\alpha_{31}\alpha_{13})x_3 + (\alpha_{34} - \alpha_{11}^{-1}\alpha_{31}\alpha_{14})x_4$$
$$= b_3 - \alpha_{11}^{-1}\alpha_{31}b_1 \qquad (12.4)$$
$$(\alpha_{42} - \alpha_{11}^{-1}\alpha_{41}\alpha_{12})x_2 + (\alpha_{43} - \alpha_{11}^{-1}\alpha_{41}\alpha_{13})x_3 + (\alpha_{44} - \alpha_{11}^{-1}\alpha_{41}\alpha_{14})x_4$$
$$= b_4 - \alpha_{11}^{-1}\alpha_{41}b_1$$

In this way we have reduced our system from one of order four to one of order three. This process is called *Gaussian elimination*.

Now if we compare the elements $\alpha_{ij} - \alpha_{11}^{-1}\alpha_{i1}\alpha_{1j}$ of the matrix $\hat{A}$ produced by the Cholesky algorithm with the coefficients of the system (12.4), we see that they are the same. In other words, Gaussian elimination and the Cholesky algorithm produce the same submatrix, and to that extent are equivalent. This is no coincidence: many direct algorithms for solving linear systems turn out to be variants of Gaussian elimination.

10. Let us now turn to the coding of Cholesky's algorithm. There are two ways to save time and storage.

1. Since A and $\hat{A}_*$ are symmetric, it is unnecessary to work with the lower half — all the information we need is in the upper half. The same applies to the other submatrices generated by the algorithm.
2. Once α and a^{T} have been used to compute ρ and r^{T}, they are no longer needed. Hence their locations can be used to store ρ and r^{T}. As the algorithm proceeds, the matrix R will overwrite the upper half of A row by row.

11. The overwriting of A by R is standard procedure, dating from the time when storage was dear and to be conserved at all costs. Perhaps now that storage is bounteous, people will quit evicting A and give R a home of its own. Time will tell.

12. The algorithm proceeds in n stages. At the first stage, the first row of R is computed and the $(n-1) \times (n-1)$ matrix A_* in the southeast corner is modified. At the second stage, the second row of R is computed and the $(n-2) \times (n-2)$ matrix in the southeast corner is modified. The process continues until it falls out of the southeast corner. Thus the algorithm begins with a loop on the row of R to be computed.

```
do 40 k=1,n
```

At the beginning of the kth stage the array that contained A has the form illustrated below for $n = 6$ and $k = 3$:

$$\begin{pmatrix} \rho_{11} & \rho_{12} & \rho_{13} & \rho_{14} & \rho_{15} & \rho_{16} \\ 0 & \rho_{22} & \rho_{23} & \rho_{24} & \rho_{25} & \rho_{26} \\ 0 & 0 & \alpha_{33} & \alpha_{34} & \alpha_{35} & \alpha_{36} \\ 0 & 0 & 0 & \alpha_{44} & \alpha_{45} & \alpha_{46} \\ 0 & 0 & 0 & 0 & \alpha_{55} & \alpha_{56} \\ 0 & 0 & 0 & 0 & 0 & \alpha_{26} \end{pmatrix}.$$

The computation of the kth row of R is straightforward:

```
        a(k,k) = sqrt(a(k,k))
        do 10 j=k+1,n
            a(k,j) = a(k,j)/a(k,k)
10      continue
```

At this point the array has the form

$$\begin{pmatrix} \rho_{11} & \rho_{12} & \rho_{13} & \rho_{14} & \rho_{15} & \rho_{16} \\ 0 & \rho_{22} & \rho_{23} & \rho_{24} & \rho_{25} & \rho_{26} \\ 0 & 0 & \rho_{33} & \rho_{34} & \rho_{35} & \rho_{36} \\ 0 & 0 & 0 & \alpha_{44} & \alpha_{45} & \alpha_{46} \\ 0 & 0 & 0 & 0 & \alpha_{55} & \alpha_{56} \\ 0 & 0 & 0 & 0 & 0 & \alpha_{26} \end{pmatrix}.$$

We must now adjust the elements beginning with α_{44}. We will do it by columns.

```
        do 30 j=k+1,n
            do 20 i=k+1,j
                a(i,j) = a(i,j) - a(k,i)*a(k,j)
20          continue
30      continue
```

Finally we must finish off the loop in k.

```
40      continue
```

Here is what the algorithm looks like when the code is assembled in one place.

```
      do 40 k=1,n
         a(k,k) = sqrt(a(k,k))
         do 10 j=k+1,n
            a(k,j) = a(k,j)/a(k,k)
10       continue
         do 30 j=k+1,n
            do 20 i=k+1,j
               a(i,j) = a(i,j) - a(k,i)*a(k,j)
20          continue
30       continue
40    continue
```

Economics

13. Since our code is in FORTRAN, we have tried to preserve column orientation by modifying A_* by columns. Unfortunately, this strategy does not work. The kth row of R is stored over the kth row of A, and we must repeatedly cross a row of A in modifying A_*. The offending reference is a(k,i) in the inner loop.

```
         do 20 i=k+1,j
            a(i,j) = a(i,j) - a(k,i)*a(k,j)
20       continue
```

There is really nothing to be done about this situation, unless we are willing to provide an extra one-dimensional array — call it r. We can then store the current row of R in r and use it to adjust the current A_*. This results in the following code.

```
      do 40 k=1,n
         a(k,k) = sqrt(a(k,k))
         do 10 j=k+1,n
            a(k,j) = a(k,j)/a(k,k)
            r(j) = a(k,j)
10       continue
         do 30 j=k+1,n
            do 20 i=k+1,j
               a(i,j) = a(i,j) - r(i)*r(j)
20          continue
30       continue
40    continue
```

The chief drawback to this alternative is that it requires an extra parameter in the calling sequence for the array r.

A second possibility is to work with the lower half of the array A and compute $L = R^{\mathrm{T}}$. For then the rows of R become columns of L.

14. The integration technique of §10.11 gives an operation count for the Cholesky algorithm as follows. The expression

$$a(i,j) = a(i,j) - r(i)*r(j)$$

in the inner loop contains one addition and one multiplication. It is executed for k=k+1,j, j=k+1,n, and k=1,n. Consequently the number of additions and multiplications will be approximately

$$\int_1^n \int_k^n \int_k^j di\, dj\, dk = \frac{1}{6}n^3.$$

15. The fact that the Cholesky algorithm is an $O(n^3)$ algorithm has important consequences for the solution of linear systems. Given the Cholesky decomposition of A, we solve the linear system $Ax = b$ by solving the two triangular systems

1. $R^{\mathrm{T}}y = b$,
2. $Rx = y$.

Now a triangular system requires $\frac{1}{2}n^2$ operations to solve, and the two systems together require n^2 operations. To the extent that the operation counts reflect actual performance, we will spend more time in the Cholesky algorithm when

$$\frac{1}{6}n^3 > n^2,$$

or when $n > 6$. For somewhat larger n, the time spent solving the triangular systems is insignificant compared to the time spent computing the Cholesky decomposition. In particular, having computed the Cholesky decomposition of a matrix of moderate size, we can solve several systems having the same matrix at practically no extra cost.

16. In §10.4 we have deprecated the practice of computing a matrix inverse to solve a linear system. Now we can see why. A good way to calculate the inverse $X = (x_1\ x_2\ \cdots\ x_n)$ of a symmetric positive-definite matrix A is to compute the Cholesky decomposition and use it to solve the systems

$$Ax_j = e_j, \qquad j = 1, 2, \ldots, n,$$

where e_j is the jth column of the identity matrix. Now if these solutions are computed in the most efficient way, they require $\frac{1}{3}n^3$ additions and multiplications—twice as many as the Cholesky decomposition. Thus the invert-and-multiply approach is much more expensive than using the decomposition directly to solve the linear system.

Lecture 13

Linear Equations

Inner-Product Form of the Cholesky Algorithm
Gaussian Elimination

Inner-product form of the Cholesky algorithm

1. The version of the Cholesky algorithm just described is sometimes called the outer-product form of the algorithm because the expression $\hat{A}_* = A_* - aa^\mathrm{T}$ involves the *outer product* aa^T. We are now going to describe a form whose computational unit is the inner product.

2. The inner-product form of the algorithm successively computes the Cholesky decompositions of the leading principal submatrices

$$A_1 = (\alpha_{11}), \quad A_2 = \begin{pmatrix} \alpha_{11} & \alpha_{12} \\ \alpha_{21} & \alpha_{22} \end{pmatrix}, \quad A_3 = \begin{pmatrix} \alpha_{11} & \alpha_{12} & \alpha_{13} \\ \alpha_{21} & \alpha_{22} & \alpha_{23} \\ \alpha_{31} & \alpha_{32} & \alpha_{33} \end{pmatrix}, \quad \ldots.$$

To get things started, note that the Cholesky factor of A_1 is the scalar $\rho_{11} = \sqrt{\alpha_{11}}$.

Now assume that we have computed the Cholesky factor R_{k-1} of A_{k-1}, and we wish to compute the Cholesky factor R_k of A_k. Partition the equation $A_k = R_k^\mathrm{T} R_k$ in the form

$$\begin{pmatrix} A_{k-1} & a_k \\ a_k^\mathrm{T} & \alpha_{kk} \end{pmatrix} = \begin{pmatrix} R_{k-1}^\mathrm{T} & 0 \\ r_k^\mathrm{T} & \rho_{kk} \end{pmatrix} \begin{pmatrix} R_{k-1} & r_k \\ 0 & \rho_{kk} \end{pmatrix}.$$

This partition gives three equations:

1. $A_{k-1} = R_{k-1}^\mathrm{T} R_{k-1}$,
2. $a_k = R_{k-1}^\mathrm{T} r_k$,
3. $\alpha_{kk} = r_k^\mathrm{T} r_k + \rho_{kk}^2$.

The first equation simply confirms that R_{k-1} is the Cholesky factor of A_{k-1}. But the second and third equations can be turned into an algorithm for computing the kth column of R: namely,

1. Solve $R_{k-1}^\mathrm{T} r_k = a_k$
2. $\rho_{kk} = \sqrt{\alpha_{kk} - r_k^\mathrm{T} r_k}$

Since R_{k-1}^T is a lower triangular matrix, the system in the first step can be easily solved. Moreover, since A_k is positive definite, ρ_{kk} must exist; i.e., $\alpha_{kk} - r_k^\mathrm{T} r_k$

must be greater than zero, so that we can take its square root. Thus we can compute the Cholesky factors of A_1, A_2, A_3, and so on until we reach $A_n = A$. The details are left as an exercise.[11]

3. The bulk of the work done by the inner-product algorithm is in the solution of the system $R_k^T r_k = a_k$, which requires $\frac{1}{2}k^2$ additions and multiplications. Since this solution step must be repeated for $k = 1, 2, \ldots, n$, the total operation count for the algorithm is $\frac{1}{6}n^3$, the same as for the outer-product form of the algorithm.

4. In fact the two algorithms not only have the same operation count, they perform the same arithmetic operations. The best way to see this is to position yourself at the (i,j)-element of the array containing A and watch what happens as the two algorithms proceed. You will find that for both algorithms the (i,j)-element is altered as follows:

$$\alpha_{ij} - \rho_{1i}\rho_{1j},$$
$$\alpha_{ij} - \rho_{1i}\rho_{1j} - \rho_{2i}\rho_{2j},$$
$$\cdots$$
$$\alpha_{ij} - \rho_{1i}\rho_{1j} - \rho_{2i}\rho_{2j} - \cdots - \rho_{i-1,i}\rho_{i-1,j}.$$

Then, depending on whether or not $i = j$, the square root of the element will be taken to give ρ_{ii}, or the element will be divided by ρ_{ii} to give ρ_{ij}.

One consequence of this observation is that the two algorithms are the same with respect to rounding errors: they give the same answers to the very last bit.

Gaussian elimination

5. We will now turn to the general nonsymmetric system of linear equations $Ax = b$. Here A is to be factored into the product $A = LU$ of a lower triangular matrix and an upper triangular matrix. The approach used to derive the Cholesky algorithm works equally well with nonsymmetric matrices; here, however, we will take another line that suggests important generalizations.

6. To motivate the approach, consider the linear system

$$\alpha_{11}x_1 + \alpha_{12}x_2 + \alpha_{13}x_3 + \alpha_{14}x_4 = b_1$$
$$\alpha_{21}x_1 + \alpha_{22}x_2 + \alpha_{23}x_3 + \alpha_{24}x_4 = b_2$$
$$\alpha_{31}x_1 + \alpha_{32}x_2 + \alpha_{33}x_3 + \alpha_{34}x_4 = b_3$$
$$\alpha_{41}x_1 + \alpha_{42}x_2 + \alpha_{43}x_3 + \alpha_{44}x_4 = b_4$$

If we set

$$m_{i1} = \alpha_{i1}/\alpha_{11}, \qquad i = 2, 3, 4,$$

---

[11] If you try for a column-oriented algorithm, you will end up computing inner products; a row-oriented algorithm requires **axpy**'s.

13. Linear Equations

and subtract m_{i1} times the first equation from the ith equation ($i = 2, 3, 4$), we end up with the system

$$\begin{aligned} \alpha_{11}x_1 + \alpha_{12}x_2 + \alpha_{13}x_3 + \alpha_{14}x_4 &= b_1 \\ \alpha'_{22}x_2 + \alpha'_{23}x_3 + \alpha'_{24}x_4 &= b'_2 \\ \alpha'_{32}x_2 + \alpha'_{33}x_3 + \alpha'_{34}x_4 &= b'_3 \\ \alpha'_{42}x_2 + \alpha'_{43}x_3 + \alpha'_{44}x_4 &= b'_4 \end{aligned}$$

where

$$\alpha'_{ij} = \alpha_{ij} - m_{i1}\alpha_{1j} \quad \text{and} \quad b'_i = b_i - m_{i1}b_1.$$

Note that the variable x_1 has been eliminated from the last three equations. Because the numbers m_{i1} multiply the first equation in the elimination they are called *multipliers*.

Now set

$$m_{i2} = a'_{i2}/a'_{22}, \quad i = 3, 4,$$

and subtract m_{i2} times the second equation from the ith equation ($i = 3, 4$). The result is the system

$$\begin{aligned} \alpha_{11}x_1 + \alpha_{12}x_2 + \alpha_{13}x_3 + \alpha_{14}x_4 &= b_1 \\ \alpha'_{22}x_2 + \alpha'_{23}x_3 + \alpha'_{24}x_4 &= b'_2 \\ \alpha''_{33}x_3 + \alpha''_{34}x_4 &= b''_3 \\ \alpha''_{43}x_3 + \alpha''_{44}x_4 &= b''_4 \end{aligned}$$

where

$$\alpha''_{ij} = \alpha'_{ij} - m_{i2}\alpha'_{2j} \quad \text{and} \quad b''_i = b'_i - m_{i2}b'_2.$$

Finally set

$$m_{i3} = a''_{i3}/a''_{33}, \quad i = 4$$

and subtract m_{i3} times the third equation from the fourth equation. The result is the upper triangular system

$$\begin{aligned} \alpha_{11}x_1 + \alpha_{12}x_2 + \alpha_{13}x_3 + \alpha_{14}x_4 &= b_1 \\ \alpha'_{22}x_2 + \alpha'_{23}x_3 + \alpha'_{24}x_4 &= b'_2 \\ \alpha''_{33}x_3 + \alpha''_{34}x_4 &= b''_3 \\ \alpha'''_{44}x_4 &= b'''_4 \end{aligned} \quad (13.1)$$

where

$$\alpha'''_{ij} = \alpha''_{ij} - m_{i3}\alpha''_{3j} \quad \text{and} \quad b'''_i = b''_i - m_{i3}b''_3.$$

Since the system (13.1) is upper triangular, it can be solved by the techniques we have already discussed.

7. The algorithm we have just described for a system of order four extends in an obvious way to systems of any order. The triangularization of the system is usually called Gaussian elimination, and the solution of the resulting triangular

system is called back substitution. Although there are slicker derivations, this one has the advantage of showing the flexibility of the algorithm. For example, if some of the elements to be eliminated are zero, we can skip their elimination with a corresponding savings in operations. We will put this flexibility to use later.

8. We have not yet connected Gaussian elimination with an LU decomposition. One way is to partition the equation $A = LU$ appropriately, derive an algorithm, and observe that the algorithm is the same as the elimination algorithm we just derived. However, there is another way.

9. Let $A_1 = A$ and set

$$M_1 = \begin{pmatrix} 1 & 0 & 0 & 0 \\ -m_{21} & 1 & 0 & 0 \\ -m_{31} & 0 & 1 & 0 \\ -m_{41} & 0 & 0 & 1 \end{pmatrix},$$

where the m_{ij}'s are the multipliers defined above. Then it follows that

$$A_2 \equiv M_1 A_1$$
$$= \begin{pmatrix} 1 & 0 & 0 & 0 \\ -m_{21} & 1 & 0 & 0 \\ -m_{31} & 0 & 1 & 0 \\ -m_{41} & 0 & 0 & 1 \end{pmatrix} \begin{pmatrix} \alpha_{11} & \alpha_{12} & \alpha_{13} & \alpha_{14} \\ \alpha_{21} & \alpha_{22} & \alpha_{23} & \alpha_{24} \\ \alpha_{31} & \alpha_{32} & \alpha_{33} & \alpha_{34} \\ \alpha_{41} & \alpha_{42} & \alpha_{43} & \alpha_{44} \end{pmatrix} = \begin{pmatrix} \alpha_{11} & \alpha_{12} & \alpha_{13} & \alpha_{14} \\ 0 & \alpha'_{22} & \alpha'_{23} & \alpha'_{24} \\ 0 & \alpha'_{32} & \alpha'_{33} & \alpha'_{34} \\ 0 & \alpha'_{42} & \alpha'_{43} & \alpha'_{44} \end{pmatrix}.$$

Next set

$$M_2 = \begin{pmatrix} 1 & 0 & 0 & 0 \\ 0 & 1 & 0 & 0 \\ 0 & -m_{32} & 1 & 0 \\ 0 & -m_{42} & 0 & 1 \end{pmatrix}.$$

Then

$$A_3 \equiv M_2 A_2$$
$$= \begin{pmatrix} 1 & 0 & 0 & 0 \\ 0 & 1 & 0 & 0 \\ 0 & -m_{32} & 1 & 0 \\ 0 & -m_{42} & 0 & 1 \end{pmatrix} \begin{pmatrix} \alpha_{11} & \alpha_{12} & \alpha_{13} & \alpha_{14} \\ 0 & \alpha'_{22} & \alpha'_{23} & \alpha'_{24} \\ 0 & \alpha'_{32} & \alpha'_{33} & \alpha'_{34} \\ 0 & \alpha'_{42} & \alpha'_{43} & \alpha'_{44} \end{pmatrix} = \begin{pmatrix} \alpha_{11} & \alpha_{12} & \alpha_{13} & \alpha_{14} \\ 0 & \alpha'_{22} & \alpha'_{23} & \alpha'_{24} \\ 0 & 0 & \alpha''_{33} & \alpha''_{34} \\ 0 & 0 & \alpha''_{43} & \alpha''_{44} \end{pmatrix}.$$

Finally set

$$M_3 = \begin{pmatrix} 1 & 0 & 0 & 0 \\ 0 & 1 & 0 & 0 \\ 0 & 0 & 1 & 0 \\ 0 & 0 & -m_{43} & 1 \end{pmatrix}.$$

Then

$$U \equiv M_3 A_3$$

$$= \begin{pmatrix} 1 & 0 & 0 & 0 \\ 0 & 1 & 0 & 0 \\ 0 & 0 & 1 & 0 \\ 0 & 0 & -m_{43} & 1 \end{pmatrix} \begin{pmatrix} \alpha_{11} & \alpha_{12} & \alpha_{13} & \alpha_{14} \\ 0 & \alpha'_{22} & \alpha'_{23} & \alpha'_{24} \\ 0 & 0 & \alpha''_{33} & \alpha''_{34} \\ 0 & 0 & \alpha''_{43} & \alpha''_{44} \end{pmatrix} = \begin{pmatrix} \alpha_{11} & \alpha_{12} & \alpha_{13} & \alpha_{14} \\ 0 & \alpha'_{22} & \alpha'_{23} & \alpha'_{24} \\ 0 & 0 & \alpha''_{33} & \alpha''_{34} \\ 0 & 0 & 0 & \alpha'''_{44} \end{pmatrix}.$$

In other words the product $U = M_3 M_2 M_1 A$ is the upper triangular matrix — the system of coefficients — produced by Gaussian elimination. If we set $L = M_1^{-1} M_2^{-1} M_3^{-1}$, then

$$A = LU.$$

Moreover, since the inverse of a lower triangular matrix is lower triangular and the product of lower triangular matrices is lower triangular, L itself is lower triangular. Thus we have exhibited an LU factorization of A.

10. We have exhibited an LU factorization, but we have not yet computed it. To do that we must supply the elements of L. And here is a surprise. *The (i,j)-element of L is just the multiplier m_{ij}.*

To see this, first note that M_k^{-1} may be obtained from M_k by flipping the sign of the multipliers; e.g.,

$$M_2^{-1} = \begin{pmatrix} 1 & 0 & 0 & 0 \\ 0 & 1 & 0 & 0 \\ 0 & m_{32} & 1 & 0 \\ 0 & m_{42} & 0 & 1 \end{pmatrix}.$$

You can establish this by showing that the product is the identity.

It is now easy to verify that

$$M_2^{-1} M_3^{-1} = \begin{pmatrix} 1 & 0 & 0 & 0 \\ 0 & 1 & 0 & 0 \\ 0 & m_{32} & 1 & 0 \\ 0 & m_{42} & 0 & 1 \end{pmatrix} \begin{pmatrix} 1 & 0 & 0 & 0 \\ 0 & 1 & 0 & 0 \\ 0 & 0 & 1 & 0 \\ 0 & 0 & m_{43} & 1 \end{pmatrix} = \begin{pmatrix} 1 & 0 & 0 & 0 \\ 0 & 1 & 0 & 0 \\ 0 & m_{32} & 1 & 0 \\ 0 & m_{42} & m_{43} & 1 \end{pmatrix}$$

and

$$L = M_1^{-1} M_2^{-1} M_3^{-1}$$

$$= \begin{pmatrix} 1 & 0 & 0 & 0 \\ m_{21} & 1 & 0 & 0 \\ m_{31} & 0 & 1 & 0 \\ m_{41} & 0 & 0 & 1 \end{pmatrix} \begin{pmatrix} 1 & 0 & 0 & 0 \\ 0 & 1 & 0 & 0 \\ 0 & m_{32} & 1 & 0 \\ 0 & m_{42} & m_{43} & 1 \end{pmatrix} = \begin{pmatrix} 1 & 0 & 0 & 0 \\ m_{21} & 1 & 0 & 0 \\ m_{31} & m_{32} & 1 & 0 \\ m_{41} & m_{42} & m_{43} & 1 \end{pmatrix},$$

which is what we wanted to show.

11. Once again, the argument does not depend on the order of the system. Hence we have the following general result.

> If Gaussian elimination is performed on a matrix of order n to give an upper triangular matrix U, then $A = LU$, where L is a lower triangular matrix with ones on its diagonal. For $i > j$ the (i,j)-element of L is the multiplier m_{ij}.

Because the diagonal elements of L are one, it is said to be *unit lower triangular*.

12. The following code overwrites A with its LU factorization. When it is done, the elements of U occupy the upper half of the array containing A, including the diagonal, and the elements of L occupy the lower half, excluding the diagonal. (The diagonal elements of L are known to be one and do not have to be stored.)

```
          do 40 k=1,n
             do 10 i=k+1,n
                a(i,k) = a(i,k)/a(k,k)
10           continue
             do 30 j=k+1,n
                do 20 i=k+1,n
                   a(i,j) = a(i,j) - a(i,k)*a(k,j)
20              continue
30           continue
40        continue
```

13. An operation count for Gaussian elimination can be obtained in the usual way by integrating the loops:

$$\int_0^n \int_k^n \int_k^n di\, dj\, dk = \frac{1}{3}n^2.$$

The count is twice that of the Cholesky algorithm, which is to be expected, since we can no longer take advantage of symmetry.

Lecture 14

Linear Equations

Pivoting
BLAS
Upper Hessenberg and Tridiagonal Systems

Pivoting

1. The leading diagonal elements at each stage of Gaussian elimination play a special role: they serve as divisors in the formulas for the multipliers. Because of their pivotal role they are called — what else — *pivots*. If the pivots are all nonzero, the algorithm goes to completion, and the matrix has an LU factorization. However, if a pivot is zero the algorithm miscarries, and the matrix may or may not have an LU factorization. The two cases are illustrated by the matrix

$$\begin{pmatrix} 0 & 1 \\ 1 & 0 \end{pmatrix},$$

which does not have an LU factorization and the matrix

$$\begin{pmatrix} 0 & 1 \\ 0 & 0 \end{pmatrix} = \begin{pmatrix} 1 & 0 \\ 0 & 1 \end{pmatrix} \begin{pmatrix} 0 & 1 \\ 0 & 0 \end{pmatrix},$$

which does, but is singular. In both cases the algorithm fails.

2. In some sense the failure of the algorithm is a blessing — it tells you that something has gone wrong. A greater danger is that the algorithm will go on to completion after encountering a small pivot. The following example shows what can happen.[12]

$$A_1 = \begin{pmatrix} 0.001 & 2.000 & 3.000 \\ -1.000 & 3.712 & 4.623 \\ -2.000 & 1.072 & 5.643 \end{pmatrix},$$

$$M_1 = \begin{pmatrix} 1.000 & 0.000 & 0.000 \\ 1000. & 1.000 & 0.000 \\ 2000. & 0.000 & 1.000 \end{pmatrix},$$

$$A_2 = \begin{pmatrix} 0.001 & 2.000 & 3.000 \\ 0.000 & 2004. & 3005. \\ 0.000 & 4001. & 6006. \end{pmatrix},$$

---

[12]This example is from G. W. Stewart, *Introduction to Matrix Computations*, Academic Press, New York, 1973.

$$M_2 = \begin{pmatrix} 1.000 & 0.000 & 0.000 \\ 0.000 & 1.000 & 0.000 \\ 0.000 & -1.997 & 1.000 \end{pmatrix},$$

$$A_3 = \begin{pmatrix} 0.001 & 2.000 & 3.000 \\ 0.000 & 2004. & 3005. \\ 0.000 & 0.000 & 5.000 \end{pmatrix}.$$

The $(3,3)$-element of A_3 was produced by cancelling three significant figures in numbers that are about 6000, and it cannot have more than one figure of accuracy. In fact the true value is $5.922\ldots$.

3. As was noted earlier, by the time cancellation occurs in a computation, the computation is already dead. In our example, death occurs in the passage from A_1 to A_2, where large multiples of the first row were added to the second and third, obliterating the significant figures in their elements. To put it another way, we would have obtained the same decomposition if we had started with the matrix

$$\tilde{A}_1 = \begin{pmatrix} 0.001 & 2.000 & 3.000 \\ -1.000 & 4.000 & 5.000 \\ -2.000 & 1.000 & 6.000 \end{pmatrix}.$$

Clearly, there will be little relation between the solution of the system $A_1 x = b$ and $\tilde{A}_1 \tilde{x} = b$.

4. If we think in terms of linear systems, a cure for this problem presents itself immediately. The original system has the form

$$\begin{aligned} 0.001 x_1 + 2.000 x_2 + 3.000 x_3 &= b_1 \\ -1.000 x_1 + 3.712 x_2 + 4.623 x_3 &= b_2 \\ -2.000 x_1 + 1.072 x_2 + 5.643 x_3 &= b_3 \end{aligned}$$

If we interchange the first and third equations, we obtain an equivalent system

$$\begin{aligned} -2.000 x_1 + 1.072 x_2 + 5.643 x_3 &= b_3 \\ 0.001 x_1 + 2.000 x_2 + 3.000 x_3 &= b_1 \\ -1.000 x_1 + 3.712 x_2 + 4.623 x_3 &= b_2 \end{aligned}$$

whose matrix

$$\hat{A}_1 = \begin{pmatrix} -2.000 & 1.072 & 5.643 \\ 0.001 & 2.000 & 3.000 \\ -1.000 & 3.712 & 4.623 \end{pmatrix}$$

can be reduced to triangular form without difficulty.

5. This suggests the following supplement to Gaussian elimination for computing the LU decomposition.

14. Linear Equations

> At the kth stage of the elimination, determine an index p_k ($k \leq p_k \leq n$) for which $|\alpha_{p_k k}|$ is largest and interchange rows k and p_k before proceeding with the elimination.

This strategy is called *partial pivoting*.[13] It insures that the multipliers are not greater than one in magnitude, since we divide by the largest element in the pivot column. Thus a gross breakdown of the kind illustrated in §14.2 cannot occur in the course of the elimination.

6. We have motivated the idea of partial pivoting by considering a system of equations; however, it is useful to have a pure matrix formulation of the process. The problem is to describe the interchange of rows in matrix terms. The following easily verified result does the job.

> Let P denote the matrix obtained by interchanging rows k and p of the identity matrix. Then PA is the matrix obtained by interchanging rows k and p of A.

The matrix P is called a (k, p) *elementary permutation*.

7. To describe Gaussian elimination with partial pivoting in terms of matrices, let P_k denote a (k, p_k) elementary permutation, and, as usual, let M_k denote the multiplier matrix. Then

$$U = M_{n-1} P_{n-1} \cdots M_2 P_2 M_1 P_1 A \tag{14.1}$$

is upper triangular.

8. Once we have decomposed A in the form (14.1), we can use the decomposition to overwrite a vector b with the solution of the linear system $Ax = b$. The algorithm goes as follows.

   1. $b = M_k P_k b$, $k = 1, 2, \ldots, n-1$
   2. $b = U^{-1} b$

9. The notation in the above algorithm must be properly interpreted. When we calculate $P_k b$, we do not form the matrix P_k and multiply b by it. Instead we retrieve the index p_k and interchange b_k with b_{p_k}. Similarly, we do not compute $M_k b$ by matrix multiplication; rather we compute

$$b_i = b_i - m_{ik} b_k, \quad i = k+1, \ldots, n.$$

Finally, the notation $U^{-1} b$ does not mean that we invert U and multiply. Instead we overwrite b with the solution of the triangular system $Uy = b$.

---

[13] There is a less frequently used strategy, called *complete pivoting*, in which both rows and columns are interchanged to bring the largest element of the matrix into the pivot position.

Notation like this is best suited for the classroom or other situations where misconceptions are easy to correct. It is risky in print, since someone will surely take it literally.

10. One drawback of the decomposition (14.1) is that it does not provide a simple factorization of the original matrix A. However, by a very simple modification of the Gaussian elimination algorithm, we can obtain an LU factorization of $P_{n-1} \cdots P_2 P_1 A$.

11. The method is best derived from a simple example. Consider A with its third and fifth rows interchanged:

$$\begin{pmatrix} a_{11} & a_{12} & a_{13} & a_{14} & a_{15} \\ a_{21} & a_{22} & a_{23} & a_{24} & a_{25} \\ a_{51} & a_{52} & a_{53} & a_{54} & a_{55} \\ a_{41} & a_{42} & a_{43} & a_{44} & a_{45} \\ a_{31} & a_{32} & a_{33} & a_{34} & a_{35} \end{pmatrix}.$$

If one step of Gaussian elimination is performed on this matrix, we get

$$\begin{pmatrix} a_{11} & a_{12} & a_{13} & a_{14} & a_{15} \\ m_{21} & a'_{22} & a'_{23} & a'_{24} & a'_{25} \\ m_{51} & a'_{52} & a'_{53} & a'_{54} & a'_{55} \\ m_{41} & a'_{42} & a'_{43} & a'_{44} & a'_{45} \\ m_{31} & a'_{32} & a'_{33} & a'_{34} & a'_{35} \end{pmatrix},$$

where the numbers m_{ij} and a'_{ij} are the same as the numbers we would have obtained by Gaussian elimination on the original matrix—after all, they are computed by the same formulas:

$$m_{i1} = m_{i1}/m_{11},$$
$$a'_{ij} = a_{ij} - m_{i1}a_{1j}.$$

If we perform a second step of Gaussian elimination, we get

$$\begin{pmatrix} a_{11} & a_{12} & a_{13} & a_{14} & a_{15} \\ m_{21} & a'_{22} & a'_{23} & a'_{24} & a'_{25} \\ m_{51} & m_{52} & a''_{53} & a''_{54} & a''_{55} \\ m_{41} & m_{42} & a''_{43} & a''_{44} & a''_{45} \\ m_{31} & m_{32} & a''_{33} & a''_{34} & a''_{35} \end{pmatrix},$$

where once again the m_{ij} and the a''_{ij} are from Gaussian elimination on the original matrix. Now note that this matrix differs from the one we would get from Gaussian elimination on the original matrix only in having its third and fifth rows interchanged. Thus if at the third step of Gaussian elimination we decide to use the fifth row as a pivot and exchange *both* the row of the submatrix and the multipliers, it would be as if we had performed Gaussian elimination without pivoting on the original matrix with its third and fifth rows interchanged.

12. This last observation is completely general.

> If in the course of Gaussian elimination with partial pivoting both the multipliers and the matrix elements are interchanged, the resulting array contains the LU decomposition of
>
> $$P_{n-1} \cdots P_2 P_1 A.$$

13. The following code implements this variant of Gaussian elimination.
```
      do 60 k=1,n-1
         maxa = abs(a(k,k))
         p(k) = k
         do 10 i=k+1,n
            if (abs(a(i,k)) .gt. maxa) then
               maxa = abs(a(i,k))
               p(k) = i
            end if
10       continue
         do 20 j=1,n
            temp = a(k,j)
            a(k,j) = a(p(k),j)                            (14.2)
            a(p(k),j) = temp
20       continue
         do 30 i=k+1,n
            a(i,k) = a(i,k)/a(k,k)
30       continue
         do 50 j=k+1,n
            do 40 i=k+1,n
               a(i,j) = a(i,j) - a(i,k)*a(k,j)
40          continue
50       continue
60    continue
```
The loop ending at 10 finds the index of the pivot row. The loop ending at 20 swaps the pivot row with row k. Note that the loop goes from 1 to n, so that the multipliers are also swapped. The rest of the code is just like Gaussian elimination without pivoting.

14. To solve the linear system $Ax = b$, note that

$$LUx = P_{n-1} \cdots P_2 P_1 Ax = P_{n-1} \cdots P_2 P_1 b.$$

Thus we first perform the interchanges on the vector b and proceed as usual to solve the two triangular systems involving L and U.

BLAS

15. Although we have recommended the BLAS for matrix computations, we have continued to code at the scalar level. The reason is that Gaussian elimination is a flexible algorithm that can be adapted to many special purposes. But to adapt it you need to know the details at the lowest level.

16. Nonetheless, the algorithm (14.2) offers many opportunities to use the BLAS. For example, the loop

```
      maxa = abs(a(k,k))
      p(k) = k
      do 10 i=k+1,n
         if (abs(a(i,k)) .gt. maxa) then
            maxa = abs(a(i,k))
            p(k) = i
         end if
10    continue
```

can be replaced with a call to a BLAS that finds the position of the largest component of the vector

$$(a(k,k), a(k+1,k), \ldots, a(n,k))^T.$$

(In the canonical BLAS, the subprogram is called imax.) The loop

```
      do 20 j=1,n
         temp = a(k,j)
         a(k,j) = a(p(k),j)
         a(p(k),j) = temp
20    continue
```

can be replaced by a call to a BLAS (swap in the canon) that swaps the vectors

$$(a(k,k), a(k,k+1), \ldots, a(k,n))$$

and

$$(a(p(k),k), a(p(k),k+1), \ldots, a(p(k),n)).$$

The loop

```
      do 30 i=k+1,n
         a(i,k) = a(i,k)/a(k,k)
30    continue
```

can be replaced by a call to a BLAS (scal) that multiplies a vector by a scalar to compute
$$\mathtt{a(k,k)}^{-1}(\mathtt{a(k,k+1)},\mathtt{a(k,k+2)},\ldots,\mathtt{a(k,n)})^\mathrm{T}.$$

17. The two inner loops

```
         do 50 j=k+1,n
            do 40 i=k+1,n
               a(i,j) = a(i,j) - a(i,k)*a(k,j)
40          continue
50       continue
```

in which most of the work is done, are the most interesting of all. The innermost loop can, of course, be replaced by an axpy that computes the vector

$$\begin{pmatrix} \mathtt{a(k+1,j)} \\ \vdots \\ \mathtt{a(n,j)} \end{pmatrix} - \mathtt{a(k,j)} \begin{pmatrix} \mathtt{a(k+1,k)} \\ \vdots \\ \mathtt{a(n,k)}. \end{pmatrix}.$$

However, the two loops together compute the difference

$$\begin{pmatrix} \mathtt{a(k+1,k+1)} & \cdots & \mathtt{a(k+1,n)} \\ \vdots & & \vdots \\ \mathtt{a(n,k+1)} & \cdots & \mathtt{a(n,n)} \end{pmatrix} - \begin{pmatrix} \mathtt{a(k+1,k)} \\ \vdots \\ \mathtt{a(n,k)} \end{pmatrix} (\mathtt{a(k,k+1)}, \ldots, \mathtt{a(k,n)}),$$

i.e., the sum of a matrix and an outer product. Since this operation occurs frequently, it is natural to assign its computation to a BLAS, which is called ger in the canon.

18. The subprogram ger is different from the BLAS we have considered so far in that it combines vectors and matrices. Since it requires $O(n^2)$ it is called a level-two BLAS.

The use of level-two BLAS can reduce the dependency of an algorithm on array orientations. For example, if we replace the loops in §14.17 with an invocation of ger, then the latter can be coded in column- or row-oriented form as required. This feature of the level-two BLAS makes the translation from FORTRAN, which is column oriented, to C, which is row oriented, much easier. It also makes it easier to write code that takes full advantage of vector supercomputers.[14]

---

[14]There is also a level-three BLAS package that performs operations between matrices. Used with a technique called *blocking,* they can increase the efficiency of some matrix algorithms, especially on vector supercomputers, but at the cost of twisting the algorithms they benefit out of their natural shape.

Upper Hessenberg and tridiagonal systems

19. As we have mentioned, Gaussian elimination can be adapted to matrices of special structure. Specifically, if an element in the pivot column is zero, we can save operations by skipping the elimination of that element. We will illustrate the technique with an upper Hessenberg matrix.

20. A matrix A is *upper Hessenberg* if
$$i > j + 1 \implies a_{ij} = 0.$$
Diagrammatically, A is upper Hessenberg if it has the form
$$A = \begin{pmatrix} X & X & X & X & X \\ X & X & X & X & X \\ 0 & X & X & X & X \\ 0 & 0 & X & X & X \\ 0 & 0 & 0 & X & X \end{pmatrix}.$$

Here we are using a convention of Jim Wilkinson in which a 0 stands for a zero element, while an X stands for an element that may or may not be zero (the presumption is that it is not).

21. From this diagrammatic form it is clear that in the first step of Gaussian elimination only the $(2,1)$-element needs to be eliminated—the rest are already zero. Thus we have only to subtract a multiple of the first row from the second to get a matrix of the form
$$\begin{pmatrix} X & X & X & X & X \\ 0 & X & X & X & X \\ 0 & X & X & X & X \\ 0 & 0 & X & X & X \\ 0 & 0 & 0 & X & X \end{pmatrix}.$$

At the second stage, we subtract a multiple of the second row from the third to get a matrix of the form
$$\begin{pmatrix} X & X & X & X & X \\ 0 & X & X & X & X \\ 0 & 0 & X & X & X \\ 0 & 0 & X & X & X \\ 0 & 0 & 0 & X & X \end{pmatrix}.$$

Two more steps of the process yield the matrices
$$\begin{pmatrix} X & X & X & X & X \\ 0 & X & X & X & X \\ 0 & 0 & X & X & X \\ 0 & 0 & 0 & X & X \\ 0 & 0 & 0 & X & X \end{pmatrix} \quad \text{and} \quad \begin{pmatrix} X & X & X & X & X \\ 0 & X & X & X & X \\ 0 & 0 & X & X & X \\ 0 & 0 & 0 & X & X \\ 0 & 0 & 0 & 0 & X \end{pmatrix},$$

14. Linear Equations   111

the last of which is upper triangular.

22. Here is code for the reduction of an upper Hessenberg matrix to triangular form. For simplicity, we leave out the pivoting. The multipliers overwrite the subdiagonal elements, and the final triangular form overwrites the matrix.

```
      do 20 k=1,n-1
         a(k+1,k) = a(k+1,k)/a(k,k)
         do 10 j=k+1,n
            a(k+1,j) = a(k+1,j) - a(k+1,k)*a(k,k+1)
  10     continue
  20  continue
```

23.   An operation count for this algorithm can be obtained as usual by integrating over the loops:

$$\int_0^n \int_k^n dj\, dk = \frac{1}{2} n^2.$$

Thus, by taking advantage of the zero elements in an upper Hessenberg matrix we have reduced its triangularization from an order n^3 process to an order n^2 process. In fact the triangularization is of the same order as the back substitution.

24.   Even greater simplifications are obtained when the matrix is *tridiagonal*, i.e., of the form

$$\begin{pmatrix} X & X & 0 & 0 & 0 \\ X & X & X & 0 & 0 \\ 0 & X & X & X & 0 \\ 0 & 0 & X & X & X \\ 0 & 0 & 0 & X & X \end{pmatrix}.$$

Here we not only reduce the number of rows to be eliminated, but (in the absence of pivoting) when two rows are combined, only the diagonal element of the second is altered, so that the computation reduces to statement

```
a(k+1,k+1) = a(k+1,k+1) - a(k+1,k)*a(k,k+1)
```

The result is an $O(n)$ algorithm. Of course, we would not waste a square array to store the $3n - 2$ numbers that represent a tridiagonal matrix. Instead we might store the diagonal elements in linear arrays. The details of this algorithm are left as an exercise.

Lecture 15

Linear Equations

Vector Norms
Matrix Norms
Relative error
Sensitivity of Linear Systems

Vector norms

1. We are going to consider the sensitivity of linear systems to errors in their coefficients. To do so, we need some way of measuring the size of the errors in the coefficients and the size of the resulting perturbation in the solution. One possibility is to report the errors individually, but for matrices this amounts to n^2 numbers — too many to examine one by one. Instead we will summarize the sizes of the errors in a single number called a *norm*. There are norms for both matrices and vectors.

2. A *vector norm* is a function $\|\cdot\| : \mathbf{R}^n \to \mathbf{R}$ that satisfies

$$
\begin{aligned}
&1. \quad x \neq 0 \implies \|x\| > 0, \\
&2. \quad \|\alpha x\| = |\alpha|\,\|x\|, \\
&3. \quad \|x + y\| \leq \|x\| + \|y\|.
\end{aligned}
\tag{15.1}
$$

The first condition says that the size of a nonzero vector is positive. The second says that if a vector is multiplied by a scalar its size changes proportionally. The third is a generalization of the fact that one side of a triangle is not greater than the sum of the other two sides: see Figure 15.1. A useful variant of the triangle inequality is

$$\|x - y\| \geq \|x\| - \|y\|.$$

3. The conditions satisfied by a vector norm are satisfied by the absolute value function on the line — in fact, the absolute value is a norm on $\mathbf{R}^1$. This means that many results in analysis can be transferred *mutatis mutandis* from the real line to $\mathbf{R}^n$.

4. Although there are infinitely many vector norms, the ones most commonly found in practice are the one-, two-, and infinity-norms. They are defined as follows:

$$
\begin{aligned}
&1. \quad \|x\|_1 = \sum_i |x_i|, \\
&2. \quad \|x\|_2 = \sqrt{\sum_i x_i^2}, \\
&\infty. \quad \|x\|_\infty = \max_i |x_i|.
\end{aligned}
$$

113

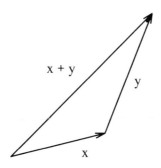

Figure 15.1. *The triangle inequality.*

In $\mathbf{R}^3$ the two-norm of a vector is its Euclidean length. Hence the two-norm is also called the *Euclidean norm*. The one norm is sometimes called the Manhattan norm, because it is the distance you have to travel to get from A to B in Manhattan (think about it). The infinity-norm is also called the max norm. All three norms are easy to compute. They satisfy

$$\|x\|_1 \geq \|x\|_2 \geq \|x\|_\infty.$$

Matrix norms

5. Matrix norms are defined in analogy with vector norms. Specifically, a matrix norm is a function $\|\cdot\| : \mathbf{R}^{m \times n} \to \mathbf{R}$ that satisfies

1. $A \neq 0 \implies \|A\| > 0$,
2. $\|\alpha A\| = |\alpha|\, \|A\|$,
3. $\|A + B\| \leq \|A\| + \|B\|$.

6. The triangle inequality allows us to bound the norm of the sum of two vectors in terms of the norms of the individual vectors. To get bounds on the products of matrices, we need another property. Specifically, let $\|\cdot\|$ stand for a family of norms defined for all matrices. Then we say that $\|\cdot\|$ is *consistent* if

$$\|AB\| \leq \|A\|\|B\|,$$

whenever the product AB is defined. A vector norm $\|\cdot\|_{\mathrm{v}}$ is consistent with a matrix norm $\|\cdot\|_{\mathrm{M}}$ if $\|Ax\|_{\mathrm{v}} \leq \|A\|_{\mathrm{M}}\|x\|_{\mathrm{v}}$.

7. The requirement of consistency frustrates attempts to generalize the vector infinity-norm in a natural way. For if we define $\|A\| = \max_i |a_{ij}|$, then

$$\left\| \begin{pmatrix} 1 & 1 \\ 1 & 1 \end{pmatrix} \begin{pmatrix} 1 & 1 \\ 1 & 1 \end{pmatrix} \right\| = \left\| \begin{pmatrix} 2 & 2 \\ 2 & 2 \end{pmatrix} \right\| = 2.$$

But
$$\left\|\begin{pmatrix} 1 & 1 \\ 1 & 1 \end{pmatrix}\right\| \left\|\begin{pmatrix} 1 & 1 \\ 1 & 1 \end{pmatrix}\right\| = 1 \cdot 1 = 1.$$
This is one reason why the matrix one- and infinity-norms have complicated definitions. Here they are — along with the two-norm, which gets new name:

1. $\|A\|_1 = \max_j \sum_i |a_{ij}|$,
F. $\|A\|_F = \sqrt{\sum_{i,j} a_{ij}^2}$,
∞. $\|A\|_\infty = \max_i \sum_j |a_{ij}|$.

The norm $\|\cdot\|_F$ is called the *Frobenius norm*.[15]

The one, Frobenius, and infinity norms are consistent. When A is a vector, the one- and infinity-norms reduce to the vector one- and infinity-norms, and the Frobenius norm reduces to the vector two-norm.

Because the one-norm is obtained by summing the absolute values of the elements in each column and taking the maximum, it is sometimes called the column-sum norm. Similarly, the infinity-norm is called the row-sum norm.

Relative error

8. Just as we use the absolute value function to define the relative error in a scalar, we can use norms to define relative errors in vectors and matrices. Specifically, the relative error in y as an approximation to x is the number

$$\rho = \frac{\|y - x\|}{\|x\|}.$$

The relative error in a matrix is defined similarly.

9. For scalars there is a close relation between relative error and the number of correct digits: if the relative error in y is ρ, then x and y agree to roughly $-\log \rho$ decimal digits. This simple relation does not hold for the components of a vector, as the following example shows.

Let
$$x = \begin{pmatrix} 1.0000 \\ 0.0100 \\ 0.0001 \end{pmatrix} \quad \text{and} \quad y = \begin{pmatrix} 1.0002 \\ 0.0103 \\ 0.0002 \end{pmatrix}.$$

In the infinity-norm, the relative error in y as an approximation to x is $3 \cdot 10^{-4}$. But the relative errors in the individual components are $2 \cdot 10^{-4}$, $3 \cdot 10^{-2}$, and 1. The large component is accurate, but the smaller components are inaccurate in proportion as they are small. This is generally true of the norms we have

---
[15] We will encounter the matrix two-norm later in §17.

introduced: the relative error gives a good idea of the accuracy of the larger components but says little about small components.

10. It sometimes happens that we are given the relative error of y as an approximation to x and want the relative error of x as an approximation to y. The following result says that when the relative errors are small, the two are essentially the same.

> If
> $$\frac{\|y - x\|}{\|x\|} \leq \rho < 1, \qquad (15.2)$$
> then
> $$\frac{\|x - y\|}{\|y\|} \leq \frac{\rho}{1 - \rho}.$$

To see this, note that from (15.2), we have
$$\rho\|x\| \geq \|y - x\| \geq \|x\| - \|y\|$$
or
$$(1 - \rho)\|x\| \leq \|y\|.$$
Hence
$$\frac{\|x - y\|}{\|y\|} \leq \frac{\|y - x\|}{(1 - \rho)\|x\|} \leq \frac{\rho}{1 - \rho}.$$

If $\rho = 0.1$, then $\rho/(1 - \rho) = 0.111\ldots$, which differs insignificantly from ρ.

Sensitivity of linear systems

11. Usually the matrix of a linear system will not be known exactly. For example, the elements of the matrix may be measured. Or they may be computed with rounding error. In either case, we end up solving not the true system
$$Ax = b,$$
but a perturbed system
$$\tilde{A}\tilde{x} = b.$$
It is natural to ask how close x is to $\tilde{x}$. This is a problem in *matrix perturbation theory*. From now on, $\|\cdot\|$ will denote both a consistent matrix norm and a vector norm that is consistent with the matrix norm.[16]

12. Let $E = \tilde{A} - A$ so that
$$\tilde{A} = A + E.$$
The first order of business is to determine conditions under which $\tilde{A}$ is nonsingular.

---

[16]We could also ask about the sensitivity of the solution to perturbations in b. This is a very easy problem, which we leave as an exercise.

15. Linear Equations

> Let A be nonsingular. If
> $$\|A^{-1}E\| < 1, \qquad (15.3)$$
> then $A + E$ is nonsingular.

To establish this result, we will show that under the condition (15.3) if $x \neq 0$ then $(A+E)x \neq 0$. Since A is nonsingular $(A+E)x = A(I+A^{-1}E)x \neq 0$ if and only if $(I + A^{-1}E)x \neq 0$. But

$$\|(I + A^{-1}E)x\| = \|x - A^{-1}Ex\| \geq \|x\| - \|A^{-1}E\|\|x\| = (1 - \|A^{-1}E\|)\|x\| > 0,$$

which establishes the result.

13. We are now in a position to establish the fundamental perturbation theorem for linear systems.

> Let A be nonsingular and let $\tilde{A} = A + E$. If
> $$Ax = b \quad \text{and} \quad \tilde{A}\tilde{x} = b,$$
> where b is nonzero, then
> $$\frac{\|\tilde{x} - x\|}{\|\tilde{x}\|} \leq \|A^{-1}E\|. \qquad (15.4)$$
> If in addition
> $$\|A^{-1}E\| < 1,$$
> then $\tilde{A}$ is nonsingular and
> $$\frac{\|\tilde{x} - x\|}{\|x\|} \leq \frac{\|A^{-1}E\|}{1 - \|A^{-1}E\|}. \qquad (15.5)$$

To establish (15.4), multiply the equation $(A + E)\tilde{x} = b$ by A^{-1} to get

$$(I + A^{-1}E)\tilde{x} = A^{-1}b = x.$$

It follows that

$$x - \tilde{x} = A^{-1}E\tilde{x},$$

or on taking norms

$$\|x - \tilde{x}\| \leq \|A^{-1}E\|\|\tilde{x}\|.$$

Since $\tilde{x} \neq 0$ (because $b \neq 0$), this inequality is equivalent to (15.4).

If $\rho = \|A^{-1}E\| < 1$, the inequality (15.4) says that the relative error in x as an approximation to $\tilde{x}$ is less than one. Hence by the result in §15.10, the relative error in $\tilde{x}$ as an approximation to x is less than or equal to $\rho/(1 - \rho)$, which is just the inequality (15.5).

Lecture 16

Linear Equations

The Condition of a Linear System
Artificial Ill-Conditioning
Rounding Error and Gaussian Elimination
Comments on the Error Analysis

The condition of a linear system

1. As it stands, the inequality

$$\frac{\|\tilde{x} - x\|}{\|x\|} \leq \frac{\|A^{-1}E\|}{1 - \|A^{-1}E\|}$$

is not easy to interpret. By weakening it we can make it more intelligible.

First note that

$$\|A^{-1}E\| \leq \|A^{-1}\|\|E\| = \kappa(A)\frac{\|E\|}{\|A\|},$$

where

$$\kappa(A) = \|A\|\|A^{-1}\|.$$

If

$$\kappa(A)\frac{\|E\|}{\|A\|} < 1,$$

then we can write

$$\frac{\|\tilde{x} - x\|}{\|x\|} \leq \frac{\kappa(A)\frac{\|E\|}{\|A\|}}{1 - \kappa(A)\frac{\|E\|}{\|A\|}}.$$

2. Now let's disassemble this inequality. First note that if $\kappa(A)\|E\|/\|A\|$ is at all small, say less than 0.1, then the denominator on the right is near one and has little effect. Thus we can consider the approximate inequality

$$\frac{\|\tilde{x} - x\|}{\|x\|} \lesssim \kappa(A)\frac{\|E\|}{\|A\|}.$$

The fraction on the left,

$$\frac{\|\tilde{x} - x\|}{\|x\|},$$

is the relative error in $\tilde{x}$ as an approximation to x. The fraction on the right,

$$\frac{\|E\|}{\|A\|},$$

is the relative error in $A + E$ as an approximation to A. Thus the number $\kappa(A)$ mediates the transfer of error from the matrix A to the solution x. As usual we call such a number a condition number — in particular, $\kappa(A)$ is the *condition number of A with respect to inversion*.[17]

3. The condition number is always greater than one:

$$1 \leq \|I\| \leq \|AA^{-1}\| \leq \|A\|\|A^{-1}\| = \kappa(A).$$

This means — unfortunately — that the condition number is a magnification constant: the bound on the error is never diminished in passing from the matrix to the solution.

4. To get a feel for what the condition number means, suppose that A is rounded on a machine with rounding unit ϵ_M, so that $\tilde{a}_{ij} = a_{ij}(1 + \epsilon_{ij})$, where $|\epsilon_{ij}| \leq \epsilon_M$. Then in any of the usual matrix norms,

$$\|E\| \leq \epsilon_M \|A\|.$$

If we solve the linear system $\tilde{A}\tilde{x} = b$ without further error, we get a solution that satisfies

$$\frac{\|\tilde{x} - x\|}{\|x\|} \leq \kappa(A)\epsilon_M. \tag{16.1}$$

In particular, if $\epsilon_M = 10^{-t}$ and $\kappa(A) = 10^k$, the solution $\tilde{x}$ can have relative error as large as 10^{-t+k}. Thus the larger components of the solution can be inaccurate in their $(t-k)$th significant figures (the smaller components can be much less accurate; see §15.9). This justifies the following rule of thumb. *If $\kappa(A) = 10^k$ expect to lose at least k digits in solving the system $Ax = b$.*

Artificial ill-conditioning

5. Unfortunately, the rule of thumb just stated is subject to the qualification that the matrix must in some sense be balanced. To see why consider the matrix

$$A = \begin{pmatrix} 1 & 1 \\ 1 & 2 \end{pmatrix},$$

whose inverse is

$$A^{-1} = \begin{pmatrix} 2 & -1 \\ -1 & 1 \end{pmatrix}.$$

The condition number of this matrix in the infinity-norm is nine.

Now multiply the first row of A by 10^{-4} to get the matrix

$$\hat{A} = \begin{pmatrix} 10^{-4} & 10^{-4} \\ 1 & 2 \end{pmatrix},$$

---

[17] Although it looks like you need a matrix inverse to compute the condition number, there are reliable ways of estimating it from the LU decomposition.

16. Linear Equations

whose inverse is

$$\hat{A}^{-1} = \begin{pmatrix} 2 \cdot 10^4 & -1 \\ -10^4 & 1 \end{pmatrix}.$$

The condition number of $\hat{A}$ is about $6 \cdot 10^4$.

If we now introduce errors into the fifth digits of the elements of A and $\hat{A}$—such errors as might be generated by rounding to four places—the infinity-norms of the error matrices will be about $\|A\|_\infty \cdot 10^{-4} = \|\hat{A}\|_\infty \cdot 10^{-4}$. Thus, for A, the error in the solution of $Ax = b$ is approximately

$$\kappa(A) \frac{\|E\|_\infty}{\|A\|_\infty} = 9 \cdot 10^{-4},$$

while for $\hat{A}$ the predicted error is

$$\kappa(\hat{A}) \frac{\|\hat{E}\|_\infty}{\|\hat{A}\|_\infty} = 6.$$

Thus we predict a small error for A and a large one for $\hat{A}$. Yet the passage from A to $\hat{A}$ is equivalent to multiplying the first equation in the system $Ax = b$ by 10^{-4}, an operation which should have no effect on the accuracy of the solution.

6. What's going on here? Is $\hat{A}$ ill conditioned or is it not? The answer is "It depends."

7. There is a sense in which $\hat{A}$ is ill conditioned. It has a row of order 10^{-4}, and a perturbation of order 10^{-4} can completely change that row—even make it zero. Thus the linear system $\hat{A}x = b$ is very sensitive to perturbations of order 10^{-4} in the first row, and that fact is reflected in the large condition number.

8. On the other hand, the errors we get by rounding the first row of $\hat{A}$ are *not all* of order 10^{-4}. Instead the errors are bounded by

$$10^{-4} \begin{pmatrix} 10^{-4} & 10^{-4} \\ 1 & 2 \end{pmatrix} = \begin{pmatrix} 10^{-8} & 10^{-8} \\ 10^{-4} & 2 \cdot 10^{-4} \end{pmatrix}.$$

The solution of $\hat{A}x = \hat{b}$ is insensitive to errors of this form, and hence the condition number is misleading.

9. To put it another way, the condition number of $\hat{A}$ has to be large to accommodate errors that can never occur in our particular application, a situation that is called *artificial ill-conditioning*. Unfortunately, there is no mechanical way to distinguish real from artificial ill-conditioning. When you get a large condition number, you have to go back to the original problem and take a hard look to see if it is truly sensitive to perturbations or is just badly scaled.

Rounding error and Gaussian elimination

10. One of the early triumphs of rounding-error analysis was the backward analysis of Gaussian elimination. Although the complete analysis is tedious, we can get a good idea of the techniques and results by looking at examples. We will start with a 2×2 matrix.

11. Let
$$A = \begin{pmatrix} 3.000 & 4.000 \\ 1.000 & 2.000 \end{pmatrix}.$$

If we perform Gaussian elimination on this matrix in four-digit arithmetic, the first (and only) multiplier is

$$m_{21} = \mathrm{fl}(a_{21}/a_{11}) = \mathrm{fl}(1.000/3.000) = 0.3333.$$

Note that if we define
$$\tilde{a}_{21} = 0.9999,$$
then
$$m_{21} = \tilde{a}_{21}/a_{11} = 0.9999/3.000 = 0.3333.$$

In other words, the multiplier we compute with rounding error is the same multiplier we would get by doing exact computations on A with its $(2,1)$-element slightly altered.

Let us continue the elimination. The reduced $(2,2)$-element is

$$a'_{22} = \mathrm{fl}(a_{22} - m_{21}a_{12}) = \mathrm{fl}(2.000 - 0.3333 \times 2.000) = \mathrm{fl}(2.000 - 0.6666) = 1.333.$$

If we replace a_{22} by
$$\tilde{a}_{22} = 1.9996,$$
then
$$a'_{22} = \tilde{a}_{22} - m_{21}a_{12} = 1.9996 - 0.3333 \times 2.000 = 1.9996 - 0.6666 = 1.333.$$

Once again, the computed value is the result of exact computing with slightly perturbed input.

To summarize, if we replace A by the nearby matrix
$$\tilde{A} = \begin{pmatrix} 3.0000 & 0.9999 \\ 2.0000 & 1.9996 \end{pmatrix}$$

and perform Gaussian elimination on $\tilde{A}$ without rounding, we get the same results as we did by performing Gaussian elimination with rounding on A.

12. The above example seems contrived. What is true of a 2×2 matrix may be false for a large matrix. And if the matrix is nearly singular, things might be even worse.

However, consider the 100×100 matrix

$$A = I - 0.01ee^{\mathrm{T}},$$

where $e^{\mathrm{T}} = (1, 1, \ldots, 1)$. This matrix has the form

$$\begin{pmatrix} 0.99 & -0.01 & \cdots & -0.01 \\ -0.01 & 0.99 & \cdots & -0.01 \\ \vdots & \vdots & & \vdots \\ -0.01 & -0.01 & \cdots & 0.99 \end{pmatrix}.$$

It is singular, which is about as ill conditioned as you can get.[18] On a machine with rounding unit $2.2 \cdot 10^{-16}$, I computed the LU decomposition of A and multiplied the factors to get a matrix $\tilde{A} = \mathrm{fl}(LU)$. In spite of the very large amount of computation involved, I found that

$$\|\tilde{A} - A\|_\infty = 7.4 \cdot 10^{-16},$$

which is less than four times the rounding unit.

13. Let us now turn from examples to analysis. The tedious part of the rounding-error analysis of Gaussian elimination is keeping track of the errors made at each stage of the elimination. However, you can catch the flavor of the analysis by looking at the first stage.

First we compute the multipliers:

$$m_{i1} = \mathrm{fl}(a_{i1}/a_{11}) = \frac{a_{i1}(1 + \epsilon_{i1})}{a_{11}},$$

where as usual $|\epsilon_{i1}| \leq \epsilon_M$. It follows that if we set

$$\tilde{a}_{i1} = a_{i1}(1 + \epsilon_{i1}), \qquad (16.2)$$

then

$$m_{i1} = \tilde{a}_{i1}/a_{11}. \qquad (16.3)$$

The next step is to reduce the submatrix:

$$\begin{aligned} a'_{ij} &= \mathrm{fl}(a_{ij} - m_{i1}a_{1j}) = [a_{ij} - m_{i1}a_{1j}(1 + \epsilon_{ij})](1 + \eta_{ij}) \\ &= [a_{ij} - m_{i1}a_{1j} + a_{ij}\eta_{ij} - m_{i1}a_{1j}(\epsilon_{ij} + \eta_{ij} + \epsilon_{ij}\eta_{ij})]. \end{aligned}$$

It follows that if we set

$$\tilde{a}_{ij} = a_{ij} + a_{ij}\eta_{ij} - m_{i1}a_{1j}(\epsilon_{ij} + \eta_{ij} + \epsilon_{ij}\eta_{ij}), \qquad (16.4)$$

---

[18] However, this does not keep Gaussian elimination from going to completion, since only the element $u_{100,100}$ is zero.

then
$$a'_{ij} = \tilde{a}_{ij} - m_{i1}a_{1j}. \tag{16.5}$$

Now it follows from (16.3) and (16.5) that the matrix

$$\begin{pmatrix} a_{11} & a_{12} & \cdots & a_{1n} \\ m_{21} & a'_{22} & \cdots & a'_{2n} \\ \vdots & \vdots & & \vdots \\ m_{n1} & a'_{n2} & \cdots & a'_{nn} \end{pmatrix},$$

which results from performing Gaussian elimination *with* rounding error on A, is the same as what we would get from performing Gaussian elimination *without* rounding error on

$$\tilde{A} = \begin{pmatrix} a_{11} & a_{12} & \cdots & a_{1n} \\ \tilde{a}_{21} & \tilde{a}_{22} & \cdots & \tilde{a}_{2n} \\ \vdots & \vdots & & \vdots \\ \tilde{a}_{n1} & \tilde{a}_{n2} & \cdots & \tilde{a}_{nn} \end{pmatrix}.$$

Moreover, A and $\tilde{A}$ are near each other. From (16.4) it follows that for $j > 1$

$$|\tilde{a}_{ij} - a_{ij}| \leq (|a_{ij}| + 3|m_{i1}||a_{1j}|)\epsilon_\mathrm{M}.$$

If we assume that the elimination is carried out with pivoting so that $|m_{i1}| \leq 1$ and set $\alpha = \max_{i,j} |a_{ij}|$, then the bound becomes

$$|\tilde{a}_{ij} - a_{ij}| \leq 4\alpha\epsilon_\mathrm{M}.$$

Similarly, (16.2) implies that this bound is also satisfied for $j = 1$.

14. All this gives the flavor of the backward rounding-error analysis of Gaussian elimination; however, there is much more to do to analyze the solution of a linear system. We will skip the details and go straight to the result.

> If Gaussian elimination with partial pivoting is used to solve the $n \times n$ system $Ax = b$ on a computer with rounding unit ϵ_M, the computed solution $\tilde{x}$ satisfies
>
> $$(A + E)\tilde{x} = b,$$
>
> where
>
> $$\frac{\|E\|}{\|A\|} \leq \varphi(n)\gamma\epsilon_\mathrm{M}. \tag{16.6}$$
>
> Here φ is a slowly growing function of n that depends on the norm, and γ is the ratio of the largest element encountered in the course of the elimination to the largest element of A.

Comments on the error analysis

15. The backward error analysis shows that the computed solution is the exact solution of a slightly perturbed matrix; that is, Gaussian elimination is a stable algorithm. As we have observed in §6.22 and §7.7, a backward error analysis is a powerful tool for understanding what went wrong in a computation. In particular, if the bound (16.6) is small, then the algorithm cannot be blamed for inaccuracies in the solution. Instead the responsible parties are the condition of the problem and (usually) errors in the initial data.

16. The function φ in the bound is a small power of n, say n^2. However, any mathematically rigorous φ is invariably an overestimate, and the error is usually of order n or less, depending on the application.

17. The number γ in the bound is called the *growth factor* because it measures the growth of elements during the elimination. If it is large, we can expect a large backward error. For example, the growth factor in the example of §14.2, where Gaussian elimination failed, was large compared to the rounding unit. In this light, partial pivoting can be seen as a way of limiting the growth by keeping the multipliers less than one.

18. Unfortunately, even with partial pivoting the growth factor can be on the order of 2^n. Consider, for example, the matrix

$$W = \begin{pmatrix} 1 & 0 & 0 & 0 & 1 \\ -1 & 1 & 0 & 0 & 1 \\ -1 & -1 & 1 & 0 & 1 \\ -1 & -1 & -1 & 1 & 1 \\ -1 & -1 & -1 & -1 & 1 \end{pmatrix}.$$

Gaussian elimination with partial pivoting applied to this matrix yields the following sequence of matrices.

$$\begin{pmatrix} 1 & 0 & 0 & 0 & 1 \\ 0 & 1 & 0 & 0 & 2 \\ 0 & -1 & 1 & 0 & 2 \\ 0 & -1 & -1 & 1 & 2 \\ 0 & -1 & -1 & -1 & 2 \end{pmatrix} \begin{pmatrix} 1 & 0 & 0 & 0 & 1 \\ 0 & 1 & 0 & 0 & 2 \\ 0 & 0 & 1 & 0 & 4 \\ 0 & 0 & -1 & 1 & 4 \\ 0 & 0 & -1 & -1 & 4 \end{pmatrix} \begin{pmatrix} 1 & 0 & 0 & 0 & 1 \\ 0 & 1 & 0 & 0 & 2 \\ 0 & 0 & 1 & 0 & 4 \\ 0 & 0 & 0 & 1 & 8 \\ 0 & 0 & 0 & -1 & 8 \end{pmatrix}$$

$$\begin{pmatrix} 1 & 0 & 0 & 0 & 1 \\ 0 & 1 & 0 & 0 & 2 \\ 0 & 0 & 1 & 0 & 4 \\ 0 & 0 & 0 & 1 & 8 \\ 0 & 0 & 0 & 0 & 16 \end{pmatrix}.$$

Clearly, if Gaussian elimination is performed on a matrix of order n having this form, the growth factor will be 2^{n-1}.

19. Does this mean that Gaussian elimination with partial pivoting is not to be trusted? The received opinion has been that examples like the one above occur only in numerical analysis texts: in real life there is little growth and often a decrease in the size of the elements. Recently, however, a naturally occurring example of exponential growth has been encountered — not surprisingly in a matrix that bears a family resemblance to W. Nonetheless, the received opinion stands. Gaussian elimination with partial pivoting is one of the most stable and efficient algorithms ever devised. Just be a little careful.

Lecture 17

Linear Equations

Introduction to a Project
More on Norms
The Wonderful Residual
Matrices with Known Condition Numbers
Invert and Multiply
Cramer's Rule
Submission

Introduction to a project

1. In this project we will use MATLAB to investigate the stability of three algorithms for solving linear systems. The algorithms are Gaussian elimination, invert-and-multiply, and Cramer's rule.

More on norms

2. We have mentioned the matrix two-norm in passing. Because the two-norm is expensive to compute, it is used chiefly in mathematical investigations. However, it is ideal for our experiments; and since MATLAB has a function `norm` that computes the two-norm, we will use it in this project. From now on $\|\cdot\|$ will denote the vector and matrix two-norms.

3. The matrix two-norm is defined by[19]

$$\|A\| = \max_{\|x\|=1} \|Ax\|.$$

Here is what you need to know about the two-norm for this project.

  1. The matrix two-norm of a vector is its vector two-norm.
  2. The matrix two-norm is consistent; that is, $\|AB\| \leq \|A\|\|B\|$, whenever AB is defined.
  3. $\|xy^\mathrm{T}\| = \|x\|\|y\|$.
  4. $\|\mathrm{diag}(d_1, \ldots, d_n)\| = \max_i\{|d_i|\}$.
  5. If $U^\mathrm{T}U = I$ and $V^\mathrm{T}V = I$ (we say U and V are *orthogonal*), then $\|U^\mathrm{T}AV\| = \|A\|$.

---

[19]If you find this definition confusing, think of it this way. Given a vector x of length one, the matrix A stretches or shrinks it into a vector of length $\|Ax\|$. The matrix two-norm of A is the largest amount it can stretch or shrink a vector.

All these properties are easy to prove from the definition of the two-norm, and you might want to try your hand at it. For the last property, you begin by establishing it for the vector two-norm.

With these preliminaries out of the way, we are ready to get down to business.

The wonderful residual

4. How can you tell if an algorithm for solving the linear system $Ax = b$ is stable — that is, if the computed solution $\tilde{x}$ satisfies a slightly perturbed system

$$(A+E)\tilde{x} = b, \tag{17.1}$$

where

$$\frac{\|E\|}{\|A\|} = O(\epsilon_M)?$$

One way is to have a backward rounding-error analysis, as we do for Gaussian elimination. But lacking that, how can we look at a computed solution and determine if it was computed stably?

5. One number we *cannot* trust is the relative error

$$\frac{\|\tilde{x} - x\|}{\|x\|}.$$

We have seen that even with a stable algorithm the relative error depends on the condition of the problem.

6. If $\tilde{x}$ satisfies (17.1), we can obtain a lower bound on $\|E\|$ by computing the *residual*

$$r = b - A\tilde{x}.$$

Specifically, since $0 = b - (A+E)\tilde{x} = r + E\tilde{x}$, we have $\|r\| \leq \|E\tilde{x}\| \leq \|E\|\|\tilde{x}\|$. It follows that

$$\frac{\|E\|}{\|A\|} \geq \frac{\|r\|}{\|A\|\|\tilde{x}\|}.$$

Thus if the *relative residual*

$$\frac{r}{\|A\|\|\tilde{x}\|}$$

has a large norm, we know that the solution was not computed stably.

On the other hand, if the relative residual is small, the result was computed stably. To see this, we must show that there is a small matrix E such that $(A+E)\tilde{x} = b$. Let

$$E = \frac{r\tilde{x}^T}{\|\tilde{x}\|^2}.$$

17. Linear Equations

Then

$$b - (A+E)\tilde{x} = (b - A\tilde{x}) - E\tilde{x} = r - \frac{r\tilde{x}^\mathrm{T}\tilde{x}}{\|\tilde{x}\|^2} = r - \frac{r\|\tilde{x}\|^2}{\|\tilde{x}\|^2} = 0,$$

so that $(A+E)\tilde{x} = b$. But

$$\frac{\|E\|}{\|A\|} = \frac{\|r\tilde{x}^\mathrm{T}\|}{\|\tilde{x}\|^2 \|A\|},$$

and it is easy to see that $\|r\tilde{x}^\mathrm{T}\| = \|r\|\|\tilde{x}\|$. Hence,

$$\frac{\|E\|}{\|A\|} = \frac{\|r\|}{\|A\|\|\tilde{x}\|}.$$

7. What we have shown is that the *relative residual norm*

$$\frac{\|r\|}{\|A\|\|\tilde{x}\|}$$

is a reliable indication of stability. A stable algorithm will yield a relative residual norm that is of the order of the rounding unit; an unstable algorithm will yield a larger value.

Matrices with known condition numbers

8. To investigate the effects of conditioning, we need to be able to generate nontrivial matrices of known condition number. Given an order n and a condition number κ we will take A in the form

$$A = UDV^\mathrm{T},$$

where U and V are random orthogonal matrices (i.e., random matrices satisfying $U^\mathrm{T}U = V^\mathrm{T}V = I$), and

$$D = \mathrm{diag}(1, \kappa^{-\frac{1}{n-1}}, \kappa^{-\frac{2}{n-1}}, \ldots, \kappa^{-1}).$$

The fact that the condition number of A is κ follows directly from the properties of the two-norm enumerated in §17.3.

9. The first part of the project is to write a function

```
function a = condmat(n, kappa)
```

to generate a matrix of order n with condition number κ. To obtain a random orthogonal matrix, use the MATLAB function rand to generate a random, normally distributed matrix. Then use the function qr to factor the random matrix into the product QR of an orthogonal matrix and an upper triangular matrix, and take Q for the random orthogonal matrix.

You can check the condition of the matrix you generate by using the function cond.

Invert and multiply

10. The purpose here is to compare the stability of Gaussian elimination with the invert-and-multiply algorithm for solving $Ax = b$. Write a function

    function invmult(n, kap)

where n is the order of the matrix A and kap is a vector of condition numbers. For each component kap(i), the function should do the following.

1. Generate a random $n \times n$ matrix A of condition kap(i).
2. Generate a (normally distributed) random n-vector x.
3. Calculate $b = Ax$.
4. Calculate the solution of the system $Ax = b$ by Gaussian elimination.
5. Calculate the solution of the system $Ax = b$ by inverting A and multiplying b by the inverse.
6. Print

    [i, kap(i); reg, rrg; rei, rri]

    where

    reg is the relative error in the solution by Gaussian elimination,
    rrg is the relative residual norm for Gaussian elimination,
    rei is the relative error in the invert-and-multiply solution,
    rri is the relative residual norm for invert-and-multiply.

11. The MATLAB left divide operator "\" is implemented by Gaussian elimination. To invert a matrix, use the function inv.

Cramer's rule

12. The purpose here is to compare the stability of Gaussian elimination with Cramer's rule for solving the 2×2 system $Ax = b$. For such a system, Cramer's rule can be written in the form

$$x_1 = (b_1 a_{22} - b_2 a_{12})/d,$$
$$x_2 = (b_2 a_{11} - b_1 a_{21})/d,$$

where

$$d = a_{11} a_{22} - a_{21} a_{12}.$$

13. Write a function

    function cramer(kap)

where kap is a vector of condition numbers. For each component kap(i), the function should do the following.

17. Linear Equations

1. Generate a random 2×2 matrix A of condition `kap(i)`.
2. Generate a (normally distributed) random 2-vector x.
3. Calculate $b = Ax$.
4. Calculate the solution of the system $Ax = b$ by Gaussian elimination.
5. Calculate the solution of the system $Ax = b$ by Cramer's rule.
6. Print

`[i, kap(i); reg, rrg; rec, rrc]`

where

- `reg` is the relative error in the solution by Gaussian elimination,
- `rrg` is the relative residual norm for Gaussian elimination,
- `rec` is the relative error in the solution by Cramer's rule,
- `rrc` is the relative residual norm for Cramer's rule.

Submission

14. Run your programs for

$$\mathtt{kap} = (1, 10^4, 10^8, 10^{12}, 10^{16})$$

using the MATLAB command `diary` to accumulate your results in a file. Edit the diary file and at the top put a brief statement in your own words of what the results mean.

POLYNOMIAL INTERPOLATION

Lecture 18

Polynomial Interpolation

Quadratic Interpolation
Shifting
Polynomial Interpolation
Lagrange Polynomials and Existence
Uniqueness

Quadratic interpolation

1. Muller's method for finding a root of the equation $f(t) = 0$ is a three-point iteration (see §4.19). Given starting values x_0, x_1, x_2 and corresponding function values f_0, f_1, f_2, one determines a quadratic polynomial

$$p(t) = a_0 + a_1 t + a_2 t^2$$

satisfying
$$p(x_i) = f_i, \quad i = 1, 2, 3. \tag{18.1}$$

The next iterate x_3 is then taken to be the root nearest x_2 of the equation $p(t) = 0$.

2. At the time the method was presented, I suggested that it would be instructive to work through the details of its implementation. One of the details is the determination of the quadratic polynomial p satisfying (18.1), an example of *quadratic interpolation*. Since quadratic interpolation exhibits many features of the general interpolation problem in readily digestible form, we will treat it first.

3. If the equations (18.1) are written out in terms of the coefficients a_0, a_1, a_2, the result is the linear system

$$\begin{pmatrix} 1 & x_0 & x_0^2 \\ 1 & x_1 & x_1^2 \\ 1 & x_2 & x_2^2 \end{pmatrix} \begin{pmatrix} a_0 \\ a_1 \\ a_2 \end{pmatrix} = \begin{pmatrix} f_0 \\ f_1 \\ f_2 \end{pmatrix}.$$

In principle, we could find the coefficients of the interpolating polynomial by solving this system using Gaussian elimination. There are three objections to this procedure.

4. First, it is not at all clear that the matrix of the system — it is called a *Vandermonde matrix* — is nonsingular. In the quadratic case it is possible to see that it is nonsingular by performing one step of Gaussian elimination and verifying that the determinant of the resulting 2×2 system is nonzero. However, this approach breaks down in the general case.

5. A second objection is that the procedure is too expensive. This objection is not strictly applicable to the quadratic case; but in general the procedure represents an $O(n^3)$ solution to a problem which, as we will see, can be solved in $O(n^2)$ operations.

6. Another objection is that the approach can lead to ill-conditioned systems. For example, if $x_0 = 100$, $x_1 = 101$, $x_2 = 102$, then the matrix of the system is

$$V = \begin{pmatrix} 1 & 100 & 10,000 \\ 1 & 101 & 10,201 \\ 1 & 102 & 10,402 \end{pmatrix}.$$

The condition number of this system is approximately 10^8.

Now the unequal scale of the columns of V suggests that there is some artificial ill-conditioning in the problem (see §16.5) — and indeed there is. But if we rescale the system, so that its matrix assumes the form

$$\hat{V} = \begin{pmatrix} 1 & 1.00 & 1.0000 \\ 1 & 1.01 & 1.0201 \\ 1 & 1.02 & 1.0402 \end{pmatrix},$$

the condition number changes to about 10^5 — still uncomfortably large, though perhaps good enough for practical purposes. This ill-conditioning, by the way, is real and will not go away with further scaling.

Shifting

7. By rewriting the polynomial in the form

$$p(t) = b_0 + b_1(t - x_2) + b_2(t - x_2)^2,$$

we can simplify the equations and remove the ill-conditioning. Specifically, the equations for the coefficients b_i become

$$\begin{pmatrix} 1 & x_0 - x_2 & (x_0 - x_2)^2 \\ 1 & x_1 - x_2 & (x_1 - x_2)^2 \\ 1 & 0 & 0 \end{pmatrix} \begin{pmatrix} b_0 \\ b_1 \\ b_2 \end{pmatrix} = \begin{pmatrix} f_0 \\ f_1 \\ f_2 \end{pmatrix}.$$

From the third equation we have

$$b_2 = f_2,$$

from which it follows that

$$\begin{pmatrix} x_0 - x_2 & (x_0 - x_2)^2 \\ x_1 - x_2 & (x_1 - x_2)^2 \end{pmatrix} \begin{pmatrix} b_0 \\ b_1 \end{pmatrix} = \begin{pmatrix} f_0 - f_2 \\ f_1 - f_2 \end{pmatrix}.$$

For our numerical example, this equation is

$$\begin{pmatrix} -2 & 4 \\ -1 & 1 \end{pmatrix} \begin{pmatrix} b_0 \\ b_1 \end{pmatrix} = \begin{pmatrix} f_0 - f_2 \\ f_1 - f_2 \end{pmatrix},$$

which is very well conditioned.

Polynomial interpolation

8. The quadratic interpolation problem has a number of features in common with the general problem.

    1. It is of low order. High-order polynomial interpolation is rare.
    2. It was introduced to derive another numerical algorithm. Not all polynomial interpolation problems originate in this way, but many numerical algorithms require a polynomial interpolant.
    3. The appearance of the problem and the nature of its solution change with a change of basis.[20] When we posed the problem in the natural basis 1, t, t^2, we got an ill-conditioned 3×3 system. On the other hand, posing the problem in the shifted basis 1, $t - x_2$, $(t - x_2)^2$ lead to a well-conditioned 2×2 system.

9. The general polynomial interpolation problem is the following.

> Given points (x_0, f_0), (x_1, f_1), ..., (x_n, f_n), where the x_i are distinct, determine a polynomial p satisfying
>
> 1. $\deg(p) \leq n$,
> 2. $p(x_i) = f_i$, $i = 0, 1, \ldots, n$.

10. If we write p in the *natural basis* to get

$$p(t) = a_0 + a_1 t + a_2 t^2 + \cdots + a_n t^n,$$

the result is the linear system

$$\begin{pmatrix} 1 & x_0 & x_0^2 & \cdots & x_0^n \\ 1 & x_1 & x_1^2 & \cdots & x_1^n \\ \vdots & \vdots & \vdots & & \vdots \\ 1 & x_n & x_n^2 & \cdots & x_n^n \end{pmatrix} \begin{pmatrix} a_0 \\ a_1 \\ \vdots \\ a_n \end{pmatrix} = \begin{pmatrix} f_0 \\ f_1 \\ \vdots \\ f_n \end{pmatrix}. \qquad (18.2)$$

The matrix of this system is called a *Vandermonde matrix*. The direct solution of Vandermonde systems by Gaussian elimination is not recommended.

Lagrange polynomials and existence

11. The existence of the interpolating polynomial p can be established in the following way. Suppose we are given n polynomials $\ell_j(t)$ that satisfy the following conditions:

$$\ell_j(x_i) = \begin{cases} 0 & \text{if } i \neq j, \\ 1 & \text{if } i = j. \end{cases} \qquad (18.3)$$

---

[20]The term "basis" is used here in its usual sense. The space of, say, quadratic polynomials is a vector space. The functions 1, t, t^2 form a basis for that space. So do the functions 1, $t - a$, $(t - a)^2$.

Then the interpolating polynomial has the form

$$p(t) = f_0 \ell_1(t) + f_1 \ell_1(t) + \cdots + f_n \ell_n(t). \tag{18.4}$$

To establish this result, note that when the right-hand side of (18.4) is evaluated at x_i, all the terms except the ith vanish, since $\ell_j(x_i)$ vanishes for $j \neq i$. This leaves $f_i \ell_i(x_i)$, which is equal to f_i, since $\ell_i(x_i) = 1$. In equations,

$$p(x_i) = \sum_{j=0}^{n} f_j \ell_j(x_i) = f_i \ell_i(x_i) = f_i.$$

12. We must now show that polynomials ℓ_j having the properties (18.3) actually exist. For $n = 2$, they are

$$\ell_0(t) = \frac{(t-x_1)(t-x_2)}{(x_0-x_1)(x_0-x_2)}, \quad \ell_1(t) = \frac{(t-x_0)(t-x_2)}{(x_1-x_0)(x_1-x_2)},$$

$$\ell_2(t) = \frac{(t-x_0)(t-x_1)}{(x_2-x_0)(x_2-x_1)}.$$

It is easy to verify that these polynomials have the desired properties.

13. Generalizing from the quadratic case, we see that the following polynomials do the job:

$$\ell_j(t) = \prod_{\substack{i=0 \\ i \neq j}}^{i=n} \frac{t - x_i}{x_j - x_i}, \quad j = 0, \ldots, n.$$

These polynomials are called *Lagrange polynomials*.

14. One consequence of the existence theorem is that equation (18.2) has a solution for any right-hand side. In other words, *the Vandermonde matrix for $n + 1$ distinct points $x_0, \ldots, x_n$ is nonsingular.*

Uniqueness

15. To establish the uniqueness of the interpolating polynomial, we use the following result from the theory of equations.

> If a polynomial of degree n vanishes at $n + 1$ distinct points, then the polynomial is identically zero.

16. Now suppose that in addition to the polynomial p the interpolation problem has another solution q. Then $r(t) = p(t) - q(t)$ is of degree not greater than n. But since $r(x_i) = p(x_i) - q(x_i) = f_i - f_i = 0$, the polynomial r vanishes at $n + 1$ points. Hence r vanishes identically, or equivalently $p = q$.

17. The condition that $\deg(p) \leq n$ in the statement of the interpolation problem appears unnatural to some people. "Why not require the polynomial

18. Polynomial Interpolation

to be exactly of degree n?" they ask. The uniqueness of the interpolant provides an answer.

Suppose, for example, we try to interpolate three points lying on a straight line by a quadratic. Now the line itself is a linear polynomial that interpolates the points. By the uniqueness theorem, the result of the quadratic interpolation must be that same straight line. What happens, of course, is that the coefficient of t^2 comes out zero.

Lecture 19

Polynomial Interpolation

Synthetic Division
The Newton Form of the Interpolant
Evaluation
Existence and Uniqueness
Divided Differences

Synthetic division

1. The interpolation problem does not end with the determination of the interpolating polynomial. In many applications one must evaluate the polynomial at a point t. As we have seen, it requires no work at all to determine the Lagrange form of the interpolant: its coefficients are the values f_i themselves. On the other hand, the individual Lagrange polynomials are tricky to evaluate. For example, products of the form

$$(x_0 - x_i) \cdots (x_{i-1} - x_i)(x_{i+1} - x_i) \cdots (x_n - x_i)$$

can easily overflow or underflow.

2. Although the coefficients of the natural form of the interpolant

$$p(t) = a_n t^n + a_{n-1} t^{n-1} + a_{n-1} t^{n-1} + \cdots + a_1 t + a_0 \qquad (19.1)$$

are not easy to determine, the polynomial can be efficiently and stably evaluated by an algorithm called *synthetic division* or *nested evaluation*.

3. To derive the algorithm, write (19.1) in the nested form

$$p(t) = ((\cdots(((a_n)t + a_{n-1})t + a_{n-2})\cdots)t + a_1)t + a_0. \qquad (19.2)$$

(It is easy to convince yourself that (19.1) and (19.2) are the same polynomial by looking at, say, the case $n = 3$. More formally, you can prove the equality by an easy induction.) This form naturally suggests the successive evaluation

$a_n,$
$(a_n)t + a_{n-1},$
$((a_n)t + a_{n-1})t + a_{n-2},$
$\cdots$
$((\cdots((a_n)t + a_{n-1})t + a_{n-2})\cdots)t + a_1,$
$((\cdots((a_n)t + a_{n-1})t + a_{n-2})\cdots)t + a_1)t + a_0.$

At each step in this evaluation the previously calculated value is multiplied by t and added to a coefficient. This leads to the following simple algorithm.

```
p = a[n];
for (i=n-1; i>=0; i--)
    p = p*t + a[i];
```

4. Synthetic division is quite efficient, requiring only n additions and n multiplications. It is also quite stable. An elementary rounding-error analysis will show that the computed value of $p(t)$ is the exact value of a polynomial $\tilde{p}$ whose coefficients differ from those of p by relative errors on the order of the rounding unit.

The Newton form of the interpolant

5. The natural form of the interpolant is difficult to determine but easy to evaluate. The Lagrange form, on the other hand, is easy to determine but difficult to evaluate. It is natural to ask, "Is there a compromise?" The answer is, "Yes, it is the *Newton form* of the interpolant."

6. The Newton form results from choosing the basis

$$1,\ t - x_0,\ (t - x_0)(t - x_1),\ \ldots,\ (t - x_0)(t - x_1)\cdots(t - x_{n-1}), \qquad (19.3)$$

or equivalently from writing the interpolating polynomial in the form

$$\begin{aligned} p(t) = {}& c_0 + c_1(t - x_0) + c_2(t - x_0)(t - x_1) + \cdots \\ & + c_n(t - x_0)(t - x_1)\cdots(t - x_{n-1}). \end{aligned} \qquad (19.4)$$

To turn this form of the interpolant into an efficient computational tool, we must show two things: how to determine the coefficients and how to evaluate the resulting polynomial. The algorithm for evaluating $p(t)$ is a variant of synthetic division, and it will be convenient to derive it while the latter algorithm is fresh in our minds.

Evaluation

7. To derive the algorithm, first write (19.4) in nested form:

$$\begin{aligned} p(t) = ((\cdots(((c_n)(t - x_{n-1}) \\ + c_{n-1})(t - x_{n-2}) + c_{n-2})\cdots)(t - x_1) + c_1)(t - x_0) + c_0. \end{aligned}$$

From this we see that the nested Newton form has the same structure as the nested natural form. The only difference is that at each nesting the multiplier t is replaced by $(t - x_i)$. Hence we get the following algorithm.

```
p = c[n];
for (i=n-1; i>=0; i--)
    p = p*(t-x[i]) + c[i];
```

8. This algorithm requires $2n$ additions and n multiplications. It is backward stable.

Existence and uniqueness

9. The existence of the Newton form of the interpolant is not a foregone conclusion. Just because one has written down a set of $n+1$ polynomials, it does not follow that all polynomials of degree n can be expressed in terms of them. For this reason we will now establish the existence of the Newton form, before going on to show how to compute it.[21]

10. We begin by evaluating $p(x_i)$, where p is in the Newton form (19.4). Now

$$p(x_0) = c_0 + c_1(x_0 - x_0) + c_2(x_0 - x_0)(x_0 - x_1) + \cdots$$
$$+ c_n(x_0 - x_0)(x_0 - x_1)\cdots(x_0 - x_{n-1})$$
$$= c_0,$$

the last n terms vanishing because they contain the zero factor $(x_0 - x_0)$. Similarly,

$$p(x_1) = c_0 + c_1(x_1 - x_0) + c_2(x_1 - x_0)(x_1 - x_1) + \cdots$$
$$+ c_n(x_1 - x_0)(x_1 - x_1)\cdots(x_1 - x_{n-1})$$
$$= c_0 + c_1(x_1 - x_0).$$

In general, $p(x_i)$ will contain only $i+1$ nonzero terms, since the last $n-i$ terms contain the factor $(x_i - x_i)$.

It now follows from the interpolation conditions $f_i = p(x_i)$ that we must determine the coefficients c_i to satisfy

$$\begin{aligned} f_0 &= c_0, \\ f_1 &= c_0 + c_1(x_1 - x_0), \\ &\cdots \\ f_n &= c_0 + c_1(x_n - x_0) + \cdots + c_n(x_n - x_0)(x_n - x_1)\cdots(x_n - x_{n-1}). \end{aligned} \qquad (19.5)$$

This is a lower triangular system whose diagonal elements are nonzero. Hence the system is nonsingular, and there are unique coefficients c_i that satisfy the interpolation conditions.

11. The matrix of the system (19.5), namely

$$\begin{pmatrix} 1 & 0 & 0 & \cdots & 0 \\ 1 & (x_1 - x_0) & 0 & \cdots & 0 \\ 1 & (x_2 - x_0) & (x_2 - x_0)(x_2 - x_1) & \cdots & 0 \\ \vdots & \vdots & \vdots & & \vdots \\ 1 & (x_n - x_0) & (x_n - x_0)(x_n - x_1) & \cdots & (x_n - x_0)(x_n - x_1)\cdots(x_n - x_{n-1}) \end{pmatrix},$$

---

[21] The existence also follows from the fact that any sequence $\{p_i\}_{i=0}^n$ of polynomials such that p_i is *exactly* of degree i forms a basis for the set of polynomials of degree n (see §23.6). The approach taken here, though less general, has pedagogical advantages.

is analogous to the Vandermonde matrix in the sense that its (i,j)-element is the $(j-1)$th basis element evaluated at x_{i-1}. The corresponding analogue for the Lagrange basis is the identity matrix. The increasingly simple structure of these matrices is reflected in the increasing ease with which we can form the interpolating polynomials in their respective bases.

12. An interesting consequence of the triangularity of the system (19.5) is that the addition of new points to the interpolation problem does not affect the coefficients we have already computed. In other words,

c_0 is the 0-degree polynomial interpolating (x_0, f_0),

$c_0 + c_1(t - x_1)$ is the 1-degree polynomial interpolating $(x_0, f_0), (x_1, f_1)$,

$c_0 + c_1(t - x_0) + c_2(t - x_0)(t - x_1)$ is the 2-degree polynomial interpolating $(x_0, f_0), (x_1, f_1), (x_1, f_1)$,

and so on.

Divided differences

13. In principle, the triangular system (19.5) can be solved in $O(n^2)$ operations to give the coefficients of the Newton interpolant. Unfortunately, the coefficients of this system can easily overflow or underflow. However, by taking a different view of the problem, we can derive a substantially different algorithm that will determine the coefficients in $O(n^2)$ operations.

14. We begin by defining the *divided difference* $f[x_0, x_2, \ldots, x_k]$ to be the coefficient of x^k in the polynomial interpolating $(x_0, f_0), (x_1, f_1), \ldots, (x_k, f_k)$. From the observations in §19.12, it follows that

$$f[x_0, x_1, \ldots, x_k] = c_k;$$

i.e., $f[x_0, x_1, \ldots, x_k]$ is the coefficient of $(t - x_0)(t - x_1) \cdots (t - x_{k-1})$ in the Newton form of the interpolant.

15. From the first equation in the system (19.5), we find that

$$f[x_0] = f_0,$$

and from the second

$$f[x_0, x_1] = \frac{f_1 - a_0}{x_1 - x_0} = \frac{f[x_1] - f[x_0]}{x_1 - x_0}.$$

Thus the first divided difference is obtained from zeroth-order divided differences by subtracting and dividing, which is why it is called a divided difference.

19. Polynomial Interpolation

16. The above expression for $f[x_0, x_1]$ is a special case of a more general relation:
$$f[x_0, x_1, \ldots, x_k] = \frac{f[x_1, x_2, \ldots, x_k] - f[x_0, x_1, \ldots, x_{k-1}]}{x_k - x_0}. \tag{19.6}$$

To establish it, let

- p be the polynomial interpolating $(x_0, f_0), (x_1, f_1), (x_k, f_k)$,
- q be the polynomial interpolating $(x_0, f_0), (x_1, f_1), (x_{k-1}, f_{k-1})$, (19.7)
- r be the polynomial interpolating $(x_1, f_1), (x_2, f_2), (x_k, f_k)$.

Then p, q, and r have leading coefficients $f[x_0, x_1, \ldots, x_k]$, $f[x_0, x_1, \ldots, x_{k-1}]$, and $f[x_1, x_2, \ldots, x_k]$, respectively.

Now we claim that
$$p(t) = q(t) + \frac{t - x_0}{x_k - x_0}[r(t) - q(t)]. \tag{19.8}$$

To see this, we will show that the left-hand side evaluates to $f_i = p(x_i)$ when $t = x_i$ $(i = 0, \ldots, k)$. The claim then follows by the uniqueness of interpolating polynomials.

For $t = x_0$,
$$q(x_0) + \frac{x_0 - x_0}{x_k - x_0}[r(x_0) - q(x_0)] = f_0,$$
since q interpolates (x_0, f_0). For $t = x_i$ $(i = 1, \ldots, k-1)$,
$$q(x_i) + \frac{x_i - x_0}{x_k - x_0}[r(x_i) - q(x_i)] = f_i + \frac{x_i - x_k}{x_0 - x_k}(f_i - f_i) = f_i.$$

Finally, for $t = x_k$,
$$q(x_k) + \frac{x_k - x_0}{x_k - x_0}[r(x_0) - q(x_0)] = q(x_k) + [r(x_k) - q(x_k)] = r(x_k) = f_k.$$

From (19.8) and (19.7) it follows that the coefficient of x^k in p is
$$f[x_0, x_1, \ldots, x_k] = \frac{f[x_1, x_2, \ldots, x_k] - f[x_0, x_1, \ldots, x_{k-1}]}{x_k - x_0},$$
which is what we wanted to show.

17. A consequence of (19.6) is that we can compute difference quotients recursively beginning with the original f_i. Specifically, let d_{ij} denote the (i, j)-entry in the following table (the indexing begins with $d_{00} = f[x_0]$):

$$\begin{array}{llll} f_0 = f[x_0] & & & \\ f_1 = f[x_1] & f[x_0, x_1] & & \\ f_2 = f[x_2] & f[x_1, x_2] & f[x_0, x_1, x_2] & \\ f_3 = f[x_3] & f[x_2, x_3] & f[x_1, x_2, x_3] & f[x_0, x_1, x_2, x_3] \end{array} \tag{19.9}$$

Then it follows that
$$d_{ij} = \frac{d_{i,j-1} - d_{i-1,j-1}}{x_i - x_{i-j}},$$
from which the array can be built up columnwise. The diagonal elements are the coefficients of the Newton interpolant.

18. It is not necessary to store the whole two-dimensional array (19.9). The following code assumes that the f_i have been stored in the array c and overwrites the array with the coefficients of the Newton interpolant. Note that the columns of (19.9) are generated from the bottom up to avoid overwriting elements prematurely.

```
for (j=2; j<=n; j++)
   for (i=n; i>=j; i--)
      c[i] = (c[i]-c[i-1])/(x[i]-x[i-j])
```

The operation count for this algorithm is n^2 additions and $\frac{1}{2}n^2$ divisions.

Lecture 20

Polynomial Interpolation

Error in Interpolation
Error Bounds
Convergence
Chebyshev Points

Error in interpolation

1. Up to this point we have treated the ordinates f_i as arbitrary numbers. We will now shift our point of view and assume that the f_i satisfy

$$f_i = f(x_i),$$

where f is a function defined on some interval of interest. As usual we will assume that f has as many derivatives as we need.

2. Let p be the polynomial interpolating f at x_0, x_1, ..., x_n. Since polynomial interpolation is often done in the hope of finding an easily evaluated approximation to f, it is natural to look for expressions for the error

$$e(t) = f(t) - p(t).$$

In what follows, we assume that t is not equal to x_0, x_1, ..., x_n (after all, the error is zero at the points of interpolation).

3. To find an expression for the error, let $q(u)$ be the polynomial of degree $n+1$ that interpolates f at the points x_0, x_1, ..., x_n, and t. The Newton form of this interpolant is

$$q(u) = c_0 + c_1(u - x_0) + \cdots + c_n(u - x_0)\cdots(u - x_{n-1}) \\ + f[x_0, \ldots, x_n, t](u - x_0)\cdots(u - x_{n-1})(u - x_n).$$

Now (see §19.12) $c_0 + c_1(u - x_0) + \cdots + c_n(u - x_0)\cdots(u - x_{n-1})$ is just the polynomial $p(u)$. Hence if we set

$$\omega(u) = (u - x_0)\cdots(u - x_{n-1})(u - x_n),$$

we have

$$q(u) = p(u) + f[x_0, \ldots, x_n, t]\omega(u).$$

But by construction $q(t) = f(t)$. Hence

$$f(t) = p(t) + f[x_0, \ldots, x_n, t]\omega(t),$$

or
$$e(t) = f(t) - p(t) = f[x_0, \ldots, x_n, t]\omega(t), \tag{20.1}$$
which is the expression we are looking for.

4. Although (20.1) reveals an elegant relation between divided differences and the error in interpolation, it does not allow us to bound the magnitude of the error. However, we can derive another, more useful, expression by considering the function
$$\varphi(u) = f(u) - p(u) - f[x_0, \ldots, x_n, t]\omega(u).$$
Here, as above, we regard u as variable and t as fixed.

Since $p(x_i) = f(x_i)$ and $\omega(x_i) = 0$,
$$\varphi(x_i) = f(x_i) - p(x_i) - f[x_0, \ldots, x_n, t]\omega(x_i) = 0.$$
Moreover, by (20.1),
$$\varphi(t) = f(t) - p(t) - f[x_0, \ldots, x_n, t]\omega(t) = 0.$$
In other words, if I is the smallest interval containing $x_0, \ldots, x_n$ and t, then
$$\varphi(u) \text{ has at least } n+2 \text{ zeros in } I.$$
By Rolle's theorem, between each of these zeros there is a zero of $\varphi'(u)$:
$$\varphi'(u) \text{ has at least } n+1 \text{ zeros in } I.$$
Similarly,
$$\varphi''(u) \text{ has at least } n \text{ zeros in } I.$$
Continuing, we find that
$$\varphi^{(n+1)}(u) \text{ has at least one zero in } I.$$
Let ξ be one of the zeros of $\varphi^{(n+1)}$ lying in I.

To get our error expression, we now evaluate $\varphi^{(n+1)}$ at ξ. Since $p(u)$ is a polynomial of degree n,
$$p^{(n+1)}(\xi) = 0.$$
Since $\omega(u) = u^{n+1} + \cdots$,
$$\omega^{(n+1)}(\xi) = (n+1)!$$
Hence
$$0 = \varphi^{(n+1)}(\xi) = f^{(n+1)}(\xi) - f[x_0, \ldots, x_n, t](n+1)!,$$
or
$$f[x_0, \ldots, x_n, t] = \frac{f^{(n+1)}(\xi)}{(n+1)!}. \tag{20.2}$$
In view of (20.1), we have the following result.

20. Polynomial Interpolation

> Let p be the polynomial interpolating f at $x_0, x_1, \ldots, x_n$. Then
>
> $$f(t) - p(t) = \frac{f^{(n+1)}(\xi)}{(n+1)!}(t-x_0)\cdots(t-x_n), \qquad (20.3)$$
>
> where ξ lies in the smallest interval containing $x_0, x_1, \ldots, x_n$ and t.

5. The point $\xi_t = \xi$ is a function of t. The proof says nothing of the properties of ξ_t, other than to locate it in a certain interval. However, it is easy to show that
$$f^{(n+1)}(\xi_t) \text{ is continuous.}$$
Just apply l'Hôpital's rule to the expression
$$f^{(n+1)}(\xi_t) = (n+1)! \frac{f(t) - p(t)}{(t-x_0)\cdots(t-x_n)}$$
at the points $x_0, \ldots, x_n$.

6. A useful and informative corollary of (20.2) is the following expression.

> $$f[x_0, x_2, \ldots, x_n] = \frac{1}{n!} f^{(n)}(\eta),$$
>
> where η lies in the interval containing $x_1, x_2, \ldots, x_n$.

In particular, if the points $x_0, \ldots, x_n$ cluster about a point t, the nth difference quotient is an approximation to $f^{(n)}(t)$.

Error bounds

7. We will now show how to use (20.3) to derive error bounds in a simple case. Let $\ell(t)$ be the linear polynomial interpolating $f(t)$ at x_0 and x_1, and suppose that
$$|f''(t)| \le M$$
in some interval of interest. Then
$$|f(t) - \ell(t)| = \frac{|f''(\xi)|}{2}|(t-x_0)(t-x_1)| \le \frac{M}{2}|(t-x_0)(t-x_1)|.$$
The further treatment of this bound depends on whether t lies outside or inside the interval $[x_0, x_1]$.

8. If t lies outside $[x_0, x_1]$, we say that we are *extrapolating* the polynomial approximation to f. Since $|(t-x_0)(t-x_1)|$ quickly becomes large as t moves

away from $[x_0, x_1]$, extrapolation is a risky business. To be sure, many numerical algorithms are based on a little judicious extrapolation. But when the newspapers carry a story asserting that the population of the U.S. will become zero in 2046 and two weeks later a learned worthy announces that the world population will become infinite in 2045, you can bet that someone has been practicing unsafe extrapolation. Economists and astrologers are mighty extrapolators.

9. If t lies inside $[x_0, x_1]$, we say that we are *interpolating* the polynomial approximation to f. (Note the new sense of the term "interpolation.") In this case we can get a uniform error bound. Specifically, the function $|(t-x_0)(t-x_1)|$ attains its maximum in $[x_0, x_1]$ at the point $t = (x_0+x_1)/2$, and that maximum is $(x_1 - x_0)^2/4$. Hence

$$t \in [x_0, x_1] \implies |f(t) - \ell(t)| \leq \frac{M}{8}(x_1 - x_0)^2. \tag{20.4}$$

10. As an application of this bound, suppose that we want to compute cheap and dirty sines by storing values of the sine function at equally spaced points and using linear interpolation to compute intermediate values. The question then arises of how small the spacing h between the points must be to achieve a prescribed accuracy.

Specifically, suppose we require the approximation to be accurate to 10^{-4}. Since the absolute value of the second derivative of the sine is bounded by one, we have from (20.4)

$$|\sin t - \ell(t)| \leq \frac{h^2}{8}.$$

Thus we must take $h^2/8 \leq 10^{-4}$ or

$$h \leq .01\sqrt{8} = 0.0283\ldots.$$

Convergence

11. The method just described for approximating the sine uses many interpolants over small intervals. Another possibility is to use a single high-order interpolant to represent the function f over the entire interval of interest. Thus suppose that for each $n = 0, 1, 2, \ldots$ we choose n equally spaced points and let p_n interpolate f at these points. If the sequence of polynomials $\{p_n\}_{n=0}^{\infty}$ converges uniformly to f, we know there will be an n for which p_n will be a sufficiently accurate approximation.

20. Polynomial Interpolation

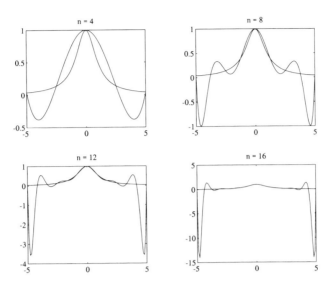

Figure 20.1. *Equally spaced interpolation.*

12. Unfortunately, equally spaced interpolation can diverge. The following example is due to Runge. Consider the function

$$f(t) = \frac{1}{1+t^2}$$

on the interval $[-5, 5]$. Figure 20.1 exhibits plots of the function and its interpolant for $n = 4$, 8, 12, and 16. You can see that the interpolants are diverging at the ends of the intervals, an observation which can be established mathematically.

Chebyshev points

13. Although it is not a complete explanation, part of the problem with Runge's example is the polynomial $\omega(t) = (t - x_0)(t - x_1) \cdots (t - x_n)$, which appears in the error expression (20.3). Figure 20.2 contains plots of ω for equally spaced points and for a set of interpolation points called Chebyshev points. The ω based on equally spaced points is the one that peaks at the ends of the interval, just about where the error in the interpolant is worst. On the other hand, the ω based on the Chebyshev points is uniformly small.

14. The Chebyshev points come from an attempt to adjust the interpolation points to control the size of ω. For definiteness, let us consider the interval

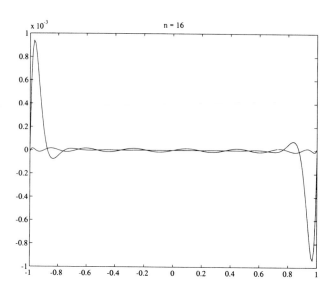

Figure 20.2. $\omega(t)$ for equally spaced and Chebyshev points.

$[-1, 1]$ and suppose that

$$|f^{(n+1)}(t)| \leq M, \quad -1 \leq t \leq 1.$$

Then from (20.3),

$$|f(t) - p_n(t)| \leq \frac{M}{n!} \max_{x \in [-1,1]} |\omega(x)|.$$

This bound will be minimized if we choose the interpolating points so that

$$\max_{x \in [-1,1]} |\omega(x)|$$

is minimized.

It can be shown

$$\min_{\omega(x)=(x-x_0)(x-x_1)\cdots(x-x_n)} \max_{x \in [-1,1]} |\omega(x)| = 2^{-n},$$

and the minimum is achieved when

$$x_i = \cos\left(\frac{2(n-i)+1}{2n+2}\pi\right), \quad i = 0, 1, \ldots, n.$$

These are the *Chebyshev points*.

20. Polynomial Interpolation

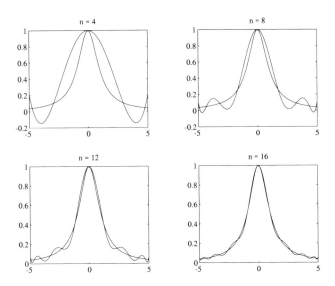

Figure 20.3. *Chebyshev interpolation.*

15. Figure 20.3 shows what happens when $1/(1+x^2)$ is interpolated at the Chebyshev points. It appears to be converging satisfactorily.

16. Unfortunately, there are functions for which interpolation at the Chebyshev points fails to converge. Moreover, better approximations of functions like $1/(1+x^2)$ can be obtained by other interpolants—e.g., cubic splines. However, if you have to interpolate a function of modest or high degree by a polynomial, you should consider basing the interpolation on the Chebyshev points.

Numerical Integration

Lecture 21

Numerical Integration

Numerical Integration
Change of Intervals
The Trapezoidal Rule
The Composite Trapezoidal Rule
Newton–Cotes Formulas
Undetermined Coefficients and Simpson's Rule

Numerical integration

1. The differential calculus is a science; the integral calculus is an art. Given a formula for a function — say e^{-x} or e^{-x^2} — it is usually possible to work your way through to its derivative. The same is not true of the integral. We can calculate

$$\int e^{-x}\, dx$$

easily enough, but

$$\int e^{-x^2}\, dx \qquad (21.1)$$

cannot be expressed in terms of the elementary algebraic and transcendental functions.

2. Sometimes it is possible to define away the problem. For example the integral (21.1) is so important in probability and statistics that there is a well-tabulated *error function*

$$\mathrm{erf}(x) = \frac{2}{\sqrt{\pi}} \int_0^x e^{-x^2}\, dx,$$

whose properties have been extensively studied. But this approach is specialized and not suitable for problems in which the function to be integrated is not known in advance.

3. One of the problems with an indefinite integral like (21.1) is that the solution has to be formula. The definite integral, on the other hand, is a number, which in principle can be computed. The process of evaluating a definite integral of a function from values of the function is called *numerical integration* or *numerical quadrature*.[22]

---

[22]The word "quadrature" refers to finding a square whose area is the same as the area under a curve.

Change of intervals

4. A typical quadrature formula is *Simpson's rule*:

$$\int_0^1 f(x)\,dx \cong \frac{1}{6}f(0) + \frac{2}{3}f\left(\frac{1}{2}\right) + \frac{1}{6}f(1).$$

Now a rule like this would not be much good if it could only be used to integrate functions over the interval $[0, 1]$. Fortunately, by performing a linear transformation of variables, we can use the rule over an arbitrary interval $[a, b]$. Since this process is used regularly in numerical integration, we describe it now.

5. The trick is to express x as a linear function of another variable y. The expression must be such that $x = a$ when $y = 0$ and $x = b$ when $y = 1$. This is a simple linear interpolation problem whose solution is

$$x = a + (b - a)y.$$

It follows that

$$dx = b - a.$$

Hence if we set

$$g(y) = f[a + (b - a)y],$$

we have

$$\int_a^b f(x)\,dx = (b - a)\int_0^1 g(y)\,dy.$$

6. For Simpson's rule we have

$$g(0) = f(a), \quad g\left(\frac{1}{2}\right) = f\left(\frac{a+b}{2}\right), \quad g(1) = f(b).$$

Hence the general form of Simpson's rule is

$$\int_a^b f(x)\,dx = \frac{b-a}{6}\left(f(a) + 4f\left(\frac{a+b}{2}\right) + f(b)\right).$$

7. This technique easily generalizes to arbitrary changes of interval, and we will silently invoke it whenever necessary.

The trapezoidal rule

8. The simplest quadrature rule in wide use is the *trapezoidal rule*. Like Newton's method, it has both a geometric and an analytic derivation. The geometric derivation is illustrated in Figure 21.1. The idea is to approximate the area under the curve $y = f(x)$ from $x = 0$ to $x = h$ by the area of the trapezoid ABCD. Now the area of a trapezoid is the product of its base with

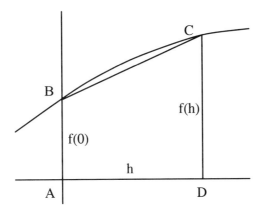

Figure 21.1. *The trapezoidal rule.*

its average height. In this case the length of the base is h, while the average height is $[f(0) + f(h)]/2$. In this way we get the trapezoidal rule

$$\int_0^h f(x)\,dx \cong \frac{h}{2}[f(0) + f(h)]. \tag{21.2}$$

9. The idea behind the analytic derivation is to interpolate $f(x)$ at 0 and h by a linear polynomial $\ell(x)$ and take $\int_0^h \ell(x)\,dx$ as an approximation to the integral of $f(x)$. The interpolant is

$$\ell(x) = f(0) + \frac{f(1) - f(0)}{h} x,$$

and an easy integration shows that $\int_0^h \ell(x)\,dx$ is the same as the right-hand side of (21.2).

10. An advantage of the analytic approach is that it leads directly to an error formula. From the theory of polynomial interpolation (see §20.4) we know that

$$f(x) - \ell(x) = \frac{f''(\xi_x)}{2} x(x - h),$$

where $\xi_x \in [0, h]$ and $f''(\xi_x)$ is a continuous function of x. If $T_h(f)$ denotes the right-hand side of (21.2), then

$$\int_0^h f(x)\,dx - T_h(f) = \int_0^h [f(x) - \ell(x)]\,dx = \frac{1}{2}\int_0^h f''(\xi_x)x(x-h)\,dx. \tag{21.3}$$

Now the function $x(x-h)$ is nonpositive on $[0, h]$. Hence by the mean value theorem for integrals, for some u and $\eta = \xi_u$, both in $[0, h]$, we have

$$\int_0^h f(x)\,dx - T_h(f) = \frac{f''(\eta)}{2}\int_0^h x(x-h)\,dx = -\frac{f''(\eta)}{12}h^3. \tag{21.4}$$

11. The error formula (21.4) shows that if $f''(x)$ is not large on $[0, h]$ and h is small, the trapezoidal rule gives a good approximation to the integral. For example if $|f''(x)| \leq 1$ and $h = 10^{-2}$, the error in the trapezoidal rule is less than 10^{-7}.

The composite trapezoidal rule

12. The trapezoidal rule cannot be expected to give accurate results over a large interval. However, by summing the results of many applications of the trapezoidal rule over smaller intervals, we can obtain an accurate approximation to the integral over any interval $[a, b]$.

13. We begin by dividing $[a, b]$ into n equal intervals by the points

$$a = x_0 < x_1 < \cdots < x_{n-1} < x_n = b.$$

Specifically, if

$$h = \frac{b-a}{n}$$

is the common length of the intervals, then

$$x_i = a + ih, \qquad i = 0, 1, \ldots, n.$$

Next we approximate $\int_{x_{i-1}}^{x_i} f(x)\,dx$ by the trapezoidal rule:

$$\int_{x_{i-1}}^{x_i} f(x)\,dx \cong \frac{h}{2}[f(x_{i-1}) + f(x_i)].$$

Finally, we sum these individual approximations to get the approximation

$$\int_a^b f(x)\,dx \cong \sum_{i=1}^n \frac{h}{2}[f(x_{i-1}) + f(x_i)].$$

After some rearranging, this sum becomes

$$\int_a^b f(x)\,dx \cong h\left(\frac{f(x_0)}{2} + f(x_1) + \cdots + f(x_{n-1}) + \frac{f(x_n)}{2}\right).$$

This formula is called the *composite trapezoidal rule*.[23]

14. We can also derive an error formula for the composite trapezoidal rule. Let $\mathrm{CT}_h(f)$ denote the approximation produced by the composite trapezoidal rule. Then from the error formula (21.4),

$$\int_a^b f(x)\,dx - \mathrm{CT}_h(f) = -\frac{h^3}{12}\sum_{i=1}^n f''(\eta_i) = -\frac{h^2}{12}\frac{(b-a)}{n}\sum_{i=1}^n f''(\eta_i),$$

---

[23] Because the composite rule is used far more often than the simple rule, people often drop the qualification "composite" and simply call it the trapezoidal rule.

21. Numerical Integration

where $\eta_i \in [x_{i-1}, x_i]$. Now the factor $\frac{1}{n}\sum_i f''(\eta_i)$ is just the arithmetic mean of the numbers $f''(\eta_i)$. Hence it lies between the largest and the smallest of these numbers, and it follows from the intermediate value theorem that there is an $\eta \in [a, b]$ such that $f''(\eta) = \frac{1}{n}\sum_i f''(\eta_i)$. Putting all this together, we get the following result.

> Let $\mathrm{CT}_h(f)$ denote the approximation produced by the composite trapezoidal rule applied to f on $[a, b]$. Then
> $$\int_a^b f(x)\,dx - \mathrm{CT}_h(f) = -\frac{(b-a)f''(\eta)}{12}h^2.$$

15. This is strong stuff. It says that we can make the approximate integral as accurate as we want simply by adding more points (compare this with polynomial interpolation where convergence cannot be guaranteed). Moreover, because the error decreases as h^2, we get twice the bang out of our added points. For example, doubling the number of points reduces the error by a factor of four.

Newton–Cotes formulas

16. From the analytic derivation of the trapezoidal rule, we see that the rule integrates any linear polynomial exactly. This suggests that we generalize the trapezoidal rule by requiring that our rule integrate exactly any polynomial of degree n. Since a polynomial of degree n has $n+1$ free parameters, it is reasonable to look for the approximate integral as a linear combination of the function evaluated at $n+1$ fixed points or abscissas. Such a quadrature rule is called a *Newton–Cotes formula*.[24]

17. Let $x_0, x_1, \ldots, x_n$ be points in the interval $[a, b]$.[25] Then we wish to determine constants $A_0, A_1, \ldots, A_n$, such that

$$\deg(f) \le n \implies \int_a^b f(x)\,dx = A_0 x_0 + A_1 x_1 + \cdots + A_n x_n. \qquad (21.5)$$

This problem has an elegant solution in terms of Lagrange polynomials.

> Let ℓ_i be the ith Lagrange polynomial over $x_0, x_1, \ldots, x_n$. Then
> $$A_i = \int_a^b \ell_i(x)\,dx \qquad (21.6)$$
> are the unique coefficients satisfying (21.5).

---

[24]Strictly speaking, the abscissas are equally spaced in a Newton–Cotes formula. But no one is making us be strict.

[25]In point of fact, the points do not have to lie in the interval, and sometimes they don't. But mostly they do.

18. To prove the assertion first note that the rule must integrate the ith Lagrange polynomial. Hence

$$\int_a^b \ell_i(x) = \sum_{j=0}^n A_j \ell_i(x_j) = A_i \ell_i(x_i) = A_i,$$

which says that the only possible value for the A_i is given by (21.6).

Now let $\deg(f) \leq n$. Then

$$f(x) = \sum_{i=0}^n f(x_i) \ell_i(x).$$

Hence

$$\int_a^b \ell_j(x)\, dx = \sum_{i=0}^n f(x_i) \int_a^b \ell_i(x)\, dx = \sum_{i=0}^n f(x_i) A_i,$$

which is just (21.5).

Undetermined coefficients and Simpson's rule

19. Although the expressions (21.6) have a certain elegance, they are difficult to evaluate. An alternative for low-order formulas is to use the exactness property (21.5) to write down a system of equations for the coefficients, a technique known as the *method of undetermined coefficients*.

20. We will illustrate the technique with a three-point formula over the interval $[0, 1]$ based on the points 0, $\frac{1}{2}$, and 1. First note that the exactness property requires that the rule integrate the function that is identically one. In other words,

$$1 \cdot A_0 + 1 \cdot A_1 + 1 \cdot A_2 = \int_0^1 1\, dx = 1.$$

The rule must also integrate the function x, which gives

$$0 \cdot A_0 + \frac{1}{2} \cdot A_1 + 1 \cdot A_2 = \int_0^1 x\, dx = \frac{1}{2}.$$

Finally, the rule must integrate the function x^2. This gives a third equation

$$0 \cdot A_0 + \frac{1}{4} \cdot A_1 + 1 \cdot A_2 = \int_0^1 x^2\, dx = \frac{1}{3}.$$

The solution of these three equations is

$$A_0 = A_2 = \frac{1}{6}$$

and

$$A_1 = \frac{2}{3}.$$

21. Numerical Integration

Thus our rule is

$$\int_0^1 f(x)\,dx \cong \frac{1}{6}f(0) + \frac{2}{3}f\left(\frac{1}{2}\right) + \frac{1}{6}f(1),$$

which is just Simpson's rule.

Lecture 22

Numerical Integration

The Composite Simpson Rule
Errors in Simpson's Rule
Treatment of Singularities
Gaussian Quadrature: The Idea

The Composite Simpson rule

1. There is a composite version of Simpson's rule for an interval $[a, b]$. To derive it, let

$$h = \frac{b-a}{n}$$

and

$$x_i = a + ih, \quad i = 0, 1, \ldots, n.$$

For brevity set

$$f_i = f(x_i).$$

Then by Simpson's rule

$$\int_{x_i}^{x_{i+2}} f(x)\, dx = \frac{h}{3}(f_i + 4f_{i+1} + f_{i+2}).$$

2. We now wish to approximate the integral of f over $[a, b]$ by summing the results of Simpson's rule over $[x_i, x_{i+2}]$. However, each application of Simpson's rule involves *two* of the intervals $[x_i, x_{i+1}]$. Thus the total number of intervals must be even. For the moment, therefore, we will assume that n is even.

3. The summation can be written as follows:

$\frac{3}{h} \int_a^b f(x)\, dx \cong f_0 + 4f_1 + f_2$
$\qquad\qquad\qquad + f_2 + 4f_3 + f_4 +$
$\qquad\qquad\qquad\qquad \cdot\ \cdot\ \cdot$
$\qquad\qquad\qquad\qquad\qquad + f_{n-4} + 4f_{n-3} + f_{n-2}$
$\qquad\qquad\qquad\qquad\qquad\qquad + f_{n-2} + 4f_{n-1} + f_n.$

This sum telescopes into

$$\int_a^b f(x)\, dx \cong \frac{h}{3}(f_0 + 4f_1 + 2f_2 + 4f_3 + 2f_4 + \cdots + 2f_{n-2} + 4f_{n-1} + f_n), \quad (22.1)$$

which is the composite Simpson rule.

4. Here is a little fragment of code that computes the sum (22.1). As above we assume that n is an even, positive integer.

```
    simp = f[0] + 4*f[1] + f[n];
    for (i=2; i<=n-2; i=i+2)
        simp = 2*f[i] + 4*f[i+1];
    simp = h*simp/3;
```

5. When n is odd, we have an extra interval $[x_{n-1}, x_n]$ over which we must integrate f. There are three possibilities.

First, we can use the trapezoidal rule to integrate over the interval. The problem with this solution is that the error in the trapezoidal rule is too big, and as h decreases it will dominate the entire sum.

Second, we can evaluate f at

$$x_{n-\frac{1}{2}} = \frac{x_{n-1} + x_n}{2}$$

and approximate the integral over $[x_{n-1}, x_n]$ by

$$\frac{h}{6}(f_{n-1} + 4f_{n-\frac{1}{2}} + f_n).$$

This solution works quite well. However, it has the drawback that it is not suitable for tabulated data, where additional function values are unobtainable.

A third option is to concoct a Newton–Cotes-type formula of the form

$$\int_{x_{n-1}}^{x_n} f(x)\,dx \cong A_0 f_{n-2} + A_1 f_{n-1} + A_2 f_n$$

and use it to integrate over the extra interval. The formula can be easily derived by the method of undetermined coefficients. It is sometimes called the *half-simp* or *semi-simp* rule.

Errors in Simpson's rule

6. It is more difficult to derive an error formula for Simpson's rule than for the trapezoidal rule. In the error formula for the trapezoidal rule (see the right-hand side of (21.3)) the polynomial $x(x - h)$ does not change sign on the interval $[0, h]$. This means that we can invoke the mean value theorem for integrals to simplify the error formula. For Simpson's rule the corresponding polynomial $x(x - h)(x - 2h)$ *does* change sign, and hence different techniques are required to get an error formula. Since these techniques are best studied in a general setting, we will just set down the error formula for Simpson's rule over the interval $[a, b]$:

$$\int_a^b f(x)\,dx - \frac{b-a}{6}\left(f(a) + 4f\left(\frac{a+b}{2}\right) + f(b)\right) = -\frac{(b-a)^5}{2880}f^{(4)}(\xi), \quad (22.2)$$

where $\xi \in [a, b]$.

7. The presence of the factor $f^{(4)}(\xi)$ on the right-hand side of (22.2) implies that the error vanishes when f is a cubic: *Although Simpson's rule was derived to be exact for quadratics, it is also exact for cubics.* This is no coincidence, as we shall see when we come to treat Gaussian quadrature.

8. The error formula for the composite Simpson rule can be obtained from (22.2) in the same way as we derived the error formula for the composite trapezoidal rule. If $\mathrm{CS}_h(f)$ denotes the result of applying the composite Simpson rule to f over the interval $[a, b]$, then

$$\int_a^b f(x)\,dx - \mathrm{CS}_h(f) = -\frac{(b-a)f^{(4)}(\xi)}{180}h^4,$$

where $\xi \in [a, b]$.

Treatment of singularities

9. It sometimes happens that one has to integrate a function with a singularity. For example, if

$$f(x) \cong \frac{c}{\sqrt{x}},$$

when x is near zero, then $\int_0^1 f(x)\,dx$ exists. However, we cannot use the trapezoidal rule or Simpson's rule to evaluate the integral because $f(0)$ is undefined.

Of course one can try to calculate the integral by a Newton–Cotes formula based on points that exclude zero; e.g., $x_0 = \frac{1}{4}$ and $x_1 = \frac{3}{4}$. However, we will still not obtain very good results, since f is not at all linear on $[0, 1]$. A better approach is to incorporate the singularity into the quadrature rule itself.

10. First define

$$g(x) = \sqrt{x} f(x).$$

Then $g(x) \cong c$ when x is near zero, so that g is well behaved. Thus we should seek a rule that evaluates the integral

$$\int_0^1 g(x) x^{-\frac{1}{2}}\,dx,$$

where g is a well-behaved function on $[0, 1]$. The function $x^{-\frac{1}{2}}$ is called a *weight function* because it assigns a weight or degree of importance to each value of the function g.

11. We will use the method of undetermined coefficients to determine a quadrature rule based on the points $x_0 = \frac{1}{4}$ and $x_1 = \frac{3}{4}$. Taking $g(x) = 1$, we get the following equation for the coefficients A_0 and A_1:

$$\int_0^1 1 \cdot x^{-\frac{1}{2}}\,dx = 2 = A_0 + A_1.$$

Taking $g(x) = x$, we get a second equation:

$$\int_0^1 x \cdot x^{-\frac{1}{2}} dx = \frac{2}{3} = \frac{1}{4}A_0 + \frac{3}{4}A_1.$$

Solving these equations, we get $A_0 = \frac{5}{3}$ and $A_1 = \frac{1}{3}$. Hence our formula is

$$\int_0^1 g(x)x^{-\frac{1}{2}} dx \cong \frac{5}{3}g\left(\frac{1}{4}\right) + \frac{2}{3}g\left(\frac{3}{4}\right).$$

12. In transforming this formula to another interval, say $[0, h]$, care must be taken to transform the weighting function properly. For example, if we wish to evaluate

$$\int_0^h g(x)x^{-\frac{1}{2}} dx,$$

we make the transformation $x = hy$, which gives

$$\int_0^h g(x)x^{-\frac{1}{2}} dx = \sqrt{h} \int_0^1 g(hx)x^{-\frac{1}{2}} dx.$$

Owing to the weight function $x^{-\frac{1}{2}}$, the transformed integral is multiplied by $\sqrt{h}$, rather than h as in the unweighted case.

13. The effectiveness of such a transformation can be seen by comparing it with an unweighted Newton–Cotes formula on the same points. The formula is easily seen to be

$$\int_0^h f(x) dx \cong \frac{h}{2}\left(f\left(\frac{h}{4}\right) + f\left(\frac{3h}{4}\right)\right).$$

The following MATLAB code compares the results of the two formulas for $h = 0.1$ applied to the function

$$\frac{\cos x}{2\sqrt{x}} - \sqrt{x} \sin x,$$

whose integral is

$$\sqrt{x} \cos x.$$

```
x = .025;
f0 = .5*cos(x)/sqrt(x) - sqrt(x)*sin(x);
g0 = .5*cos(x) - x*sin(x);
x = .075;
f1 = .5*cos(x)/sqrt(x) - sqrt(x)*sin(x);
g1 = .5*cos(x) - x*sin(x);
[.05*(f0+f1), sqrt(.1)*(5*g0/3 + g1/3), sqrt(.1)*cos(.1)]
```

The true value of the integral is 0.3146. The Newton–Cotes approximation is 0.2479. The weighted approximation is 0.3151—a great improvement.

Gaussian quadrature: The idea

14. We have observed that the coefficients A_i in the integration rule

$$\int_a^b f(x)\,dx \cong A_0 f(x_0) + A_1 f(x_1) + \cdots + A_n f(x_n)$$

represent $n+1$ degrees of freedom that we can use to make the rule exact for polynomials of degree n or less. The key idea behind Gaussian quadrature is that the abscissas $x_1, x_2, \ldots, x_n$ represent another $n+1$ degrees of freedom, which can be used to extend the exactness of the rule to polynomials of degree $2n+1$.

15. Unfortunately, any attempt to derive a Gaussian quadrature rule by the method of undetermined coefficients (and abscissas) must come to grips with the fact that the resulting equations are nonlinear. For example, when $n=1$ and $[a,b]=[0,1]$, the equations obtained by requiring the rule to integrate 1, x, x^2, and x^3 are

$$1 = A_0 + A_1$$
$$\tfrac{1}{2} = x_0 A_0 + x_1 A_1$$
$$\tfrac{1}{3} = x_0^2 A_0 + x_1^2 A_1$$
$$\tfrac{1}{4} = x_0^3 A_0 + x_1^3 A_1$$

Although this tangle of equations can be simplified, in general the approach leads to ill-conditioned systems. As an alternative, we will do as Gauss did and approach the problem through the theory of orthogonal polynomials, a theory that has wide applicability in its own right.

16. But first let's digress a bit and consider a special case. Rather than freeing all the abscissas, we could fix $n-1$ of them, allowing only one to be free. For example, to get a three-point rule that integrates cubics, we might take $x_0 = 0$ and $x_2 = 1$, leaving x_1 to be determined. This leads to the equations

$$1 = A_0 + A_1 + A_2$$
$$\tfrac{1}{2} = \phantom{A_0 +{}} x_1 A_1 + A_2$$
$$\tfrac{1}{3} = \phantom{A_0 +{}} x_1^2 A_1 + A_2$$
$$\tfrac{1}{4} = \phantom{A_0 +{}} x_1^3 A_1 + A_2$$

When these equations are solved, the result is Simpson's rule. Thus the unexpected accuracy of Simpson's rule can be explained by the fact that it is actually a constrained Gaussian quadrature formula.

Lecture 23

Numerical Integration

Gaussian Quadrature: The Setting
Orthogonal Polynomials
Existence
Zeros of Orthogonal Polynomials
Gaussian Quadrature
Error and Convergence
Examples

Gaussian quadrature: The setting

1. The Gauss formula we will actually derive has the form

$$\int_a^b f(x)w(x)\,dx \cong A_0 f(x_0) + A_1 f(x_1) + \cdots + A_n f(x_n),$$

where $w(x)$ is a weight function that is greater than zero on the interval $[a, b]$.

2. The incorporation of a weight function creates no complications in the theory. However, it makes our integrals, which are already too long, even more cumbersome. Since the interval $[a,b]$ and the weight $w(x)$ do not change, we will suppress them along with the variable of integration and write

$$\int f = \int_a^b f(x)w(x)\,dx.$$

3. Regarded as an operator on functions, $\int$ is linear. That is, $\int \alpha f = \alpha \int f$ and $\int (f+g) = \int f + \int g$. We will make extensive use of linearity in what follows.

Orthogonal polynomials

4. Two functions f and g are said to be *orthogonal* if

$$\int fg = 0.$$

The term "orthogonal" derives from the fact that the integral $\int fg$ can be regarded as an inner product of f and g. Thus two polynomials are orthogonal if their inner product is zero, which is the usual definition of orthogonality in $\mathbf{R}^n$.

5. A sequence of *orthogonal polynomials* is a sequence $\{p_i\}_{i=0}^\infty$ of polynomials with $\deg(p_i) = i$ such that

$$i \neq j \implies \int p_i p_j = 0. \tag{23.1}$$

Since orthogonality is not altered by multiplication by a nonzero constant, we may normalize the polynomial p_i so that the coefficient of x^i is one: i.e.,

$$p_i(x) = x^i + a_{i,i-1}x^{i-1} + \cdots + a_{i0}.$$

Such a polynomial is said to be *monic*.

6. Our immediate goal is to establish the existence of orthogonal polynomials. Although we could, in principle, determine the coefficients a_{ij} of p_i in the natural basis by using the orthogonality conditions (23.1), we get better results by expressing p_{n+1} in terms of lower-order orthogonal polynomials. To do this we need the following general result.

> Let $\{p_i\}_{i=0}^{\infty}$ be a sequence of polynomials such that p_i is exactly of degree i. If
> $$q(x) = a_n x^n + a_{n-1} x^{n-1} + \cdots + a_0, \qquad (23.2)$$
> then q can be written uniquely in the form
> $$q = b_n p_n + b_{n-1} p_{n-1} + \cdots + b_0 p_0. \qquad (23.3)$$

7. In establishing this result, we may assume that the polynomials p_i are monic. The proof is by induction. For $n = 0$, we have

$$q(x) = a_0 = a_0 \cdot 1 = a_0 p_0(x).$$

Hence we must have $b_0 = a_0$.

Now assume that q has the form (23.2). Since p_n is the only polynomial in the sequence $p_n, p_{n-1}, \ldots, p_0$ that contains x^n and since p_n is monic, it follows that we must have $b_n = a_n$. Then the polynomial $q - a_n p_n$ is of degree $n - 1$. Hence by the induction hypothesis, it can be expressed uniquely in the form

$$q - a_n p_n = b_{n-1} p_{n-1} + \cdots + b_0 p_0,$$

which establishes the result.

8. A consequence of this result is the following.

> The polynomial p_{n+1} is orthogonal to any polynomial q of degree n or less.

For from (23.3) it follows that

$$\int p_{n+1} q = b_n \int p_{n+1} p_n + \cdots + b_0 \int p_{n+2} p_0 = 0,$$

the last equality following from the orthogonality of the polynomials p_i.

23. Numerical Integration

Existence

9. To establish the existence of orthogonal polynomials, we begin by computing the first two. Since p_0 is monic and of degree zero,

$$p_0(x) \equiv 1.$$

Since p_1 is monic and of degree one, it must have the form

$$p_1(x) = x - \alpha_1.$$

To determine α_1, we use orthogonality:

$$0 = \int p_1 p_0 = \int (x - \alpha_1) \cdot 1 = \int x - \alpha_1 \int 1.$$

Since the function 1 is positive in the interval of integration, $\int 1 > 0$, and it follows that

$$\alpha_1 = \frac{\int x}{\int 1}.$$

10. In general we will seek p_{n+1} in the form

$$p_{n+1} = x p_n - \alpha_{n+1} p_n - \beta_{n+1} p_{n-1} - \gamma_{n+1} p_{n-2} - \cdots.$$

As in the construction of p_1, we use orthogonality to determine the coefficients $\alpha_{n+1}, \beta_{n+1}, \gamma_{n+1}, \ldots$.

To determine α_{n+1}, write

$$0 = \int p_{n+1} p_n = \int x p_n p_n - \alpha_{n+1} \int p_n p_n - \beta_{n+1} \int p_{n-1} p_n - \gamma_{n+1} \int p_{n-2} p_n - \cdots.$$

By orthogonality, $0 = \int p_{n-1} p_n = \int p_{n-2} p_n = \cdots$. Hence

$$\int x p_n^2 - \alpha_{n+1} \int p_n^2 = 0.$$

Since $\int p_n^2 > 0$, we may solve this equation to get

$$\alpha_{n+1} = \frac{\int x p_n^2}{\int p_n^2}.$$

For β_{n+1}, write

$$0 = \int p_{n+1} p_{n-1} = \int x p_n p_{n-1} - \alpha_{n+1} \int p_n p_{n-1} \\ - \beta_{n+1} \int p_{n-1} p_{n-1} - \gamma_{n+1} \int p_{n-2} p_{n-1} - \cdots.$$

Dropping terms that are zero because of orthogonality, we get

$$\int x p_n p_{n-1} - \beta_{n+1} \int p_{n-1}^2 = 0$$

or

$$\beta_{n+1} = \frac{\int x p_n p_{n-1}}{\int p_{n-1}^2}.$$

11. The formulas for the remaining coefficients are similar to the formula for β_{k+1}; e.g.,

$$\gamma_{n+1} = \frac{\int x p_n p_{n-2}}{\int p_{n-2}^2}.$$

However, there is a surprise here. The denominator $\int x p_n p_{n-2}$ can be written in the form $\int x p_{n-2} p_n$. Since $x p_{n-2}$ is of degree $n-1$ it is orthogonal to p_n; i.e., $\int x p_{n-2} p_{n-1} = 0$. Hence $\gamma_{k+1} = 0$, and likewise the coefficients of p_{n-3}, p_{n-4}, ... are zero.

12. To summarize:

> The orthogonal polynomials can be generated by the following recurrence:
>
> $$\begin{aligned} p_0 &= 1, \\ p_1 &= x - \alpha_1, \\ p_{n+1} &= x p_n - \alpha_{n+1} p_n - \beta_{n+1} p_{n-1}, \qquad n = 1, 2, \ldots, \end{aligned}$$
>
> where
>
> $$\alpha_{n+1} = \frac{\int x p_n^2}{\int p_n^2} \quad \text{and} \quad \beta_{n+1} = \frac{\int x p_n p_{n-1}}{\int p_{n-1}^2}.$$

The first two equations in the recurrence merely start things off. The right-hand side of the third equation contains three terms and for that reason is called the *three-term recurrence* for the orthogonal polynomials.

Zeros of orthogonal polynomials

13. It will turn out that the abscissas of our Gaussian quadrature formula will be the zeros of p_{n+1}. We will now show that

> The zeros of p_{n+1} are real, simple, and lie in the interval $[a, b]$.

14. Let $x_0, x_1, \ldots, x_k$ be the zeros of odd multiplicity of p_{n+1} in $[a, b]$; i.e., $x_0, x_1, \ldots, x_k$ are the points at which p_{n+1} changes sign in $[a, b]$. If $k = n$, we are through, since the x_i are the $n+1$ zeros of p_{n+1}.

Suppose then that $k < n$ and consider the polynomial

$$q(x) = (x - x_0)(x - x_1) \cdots (x - x_k).$$

Since $\deg(q) = k + 1 < n + 1$, by orthogonality

$$\int p_{n+1} q = 0.$$

23. Numerical Integration

On the other hand $p_{n+1}(x)q(x)$ cannot change sign on $[a,b]$—each sign change in $p_{n+1}(x)$ is cancelled by a corresponding sign change in $q(x)$. It follows that

$$\int p_{n+1}q \neq 0,$$

which is a contradiction.

Gaussian quadrature

15. The Gaussian quadrature formula is obtained by constructing a Newton–Cotes formula on the zeros of the orthogonal polynomial p_{n+1}.

> Let $x_0, x_1, \ldots, x_n$ be the zeros of the orthogonal polynomial p_{n+1} and set
> $$A_i = \int \ell_i, \quad i = 0, 1, \ldots, n,$$
> where ℓ_i is the ith Lagrange polynomial over $x_0, x_1, \ldots, x_n$. For any function f let
> $$G_n f = A_0 f(x) + A_1 f(x_1) + \cdots + A_n f(x_n).$$
> Then
> $$\deg(f) \leq 2n+1 \implies \int f = G_n f.$$

16. To establish this result, first note that by construction the integration formula $G_n f$ is exact for polynomials of degree less than or equal to n (see §21.17).

Now let $\deg(f) \leq 2n+1$. Divide f by p_{n+1} to get

$$f = p_{n+1}q + r, \quad \deg(q), \deg(r) \leq n. \tag{23.4}$$

Then

$$\begin{aligned}
G_n f &= \sum_i A_i f(x_i) \\
&= \sum_i A_i [p_{n+1}(x_i)q(x_i) + r(x_i)] && \text{by (23.4)} \\
&= \sum_i A_i r(x_i) && \text{because } p_{n+1}(x_i) = 0 \\
&= G_n r \\
&= \int r && \text{because } G_n \text{ is exact for } \deg(r) \leq n \\
&= \int (p_{n+1}q + r) && \text{because } \int p_{n+1}q = 0 \text{ for } \deg(q) \leq n \\
&= \int f && \text{by (23.4)}.
\end{aligned}$$

Quod erat demonstrandum.

17. An important corollary of these results is that the coefficients A_i are positive. To see this note that

$$\ell_i(x_j) = \ell_i^2(x_j) = \begin{cases} 0 & \text{if } i \neq j, \\ 1 & \text{if } i = j. \end{cases}$$

Since $\ell_i^2(x) \geq 0$ and $\deg(\ell_i^2) = 2n$,

$$0 < \textstyle\int \ell_i^2 = G_n \ell_i^2 = \sum_j A_i \ell_i^2(x_j) = A_i.$$

18. Since $A_0 + \cdots + A_n = \int 1$, no coefficient can be larger than $\int 1$. Consequently, we cannot have a situation in which large coefficients create large intermediate results that suffer cancellation when they are added.

Error and convergence

19. Gaussian quadrature has error formulas similar to the ones for Newton–Cotes formulas. Specifically

$$\textstyle\int f - G_n f = \frac{f^{(2n+2)}(\xi)}{(2n+2)!} \int p_{n+1}^2,$$

where $\xi \in [a, b]$.

20. A consequence of the positivity of the coefficients A_i is that Gaussian quadrature converges for any continuous function; that is,

$$f \text{ continuous} \implies \lim_{n \to \infty} G_n f = \textstyle\int f.$$

The proof — it is a good exercise in elementary analysis — is based on the Weierstrass approximation theorem, which says that for any continuous function f there is a sequence of polynomials that converges uniformly to f.

Examples

21. Particular Gauss formulas arise from particular choices of the interval $[a, b]$ and the weight function $w(x)$. The workhorse is *Gauss–Legendre* quadrature,[26] in which $[a, b] = [-1, 1]$ and $w(x) \equiv 1$, so that the formula approximates the integral

$$\int_{-1}^{1} f(x)\,dx.$$

The corresponding orthogonal polynomials are called Legendre polynomials.

22. If we take $[a, b] = [0, \infty]$ and $w(x) = e^{-x}$, we get a formula to approximate

$$\int_{0}^{\infty} f(x) e^{-x}\,dx.$$

This is *Gauss–Laguerre* quadrature.

---

[26]Curious bedfellows! Gauss and Legendre became involved in a famous priority dispute over the invention of least squares, and neither would enjoy seeing their names coupled this way.

23. Numerical Integration

23. If we take $[a,b] = [-\infty, \infty]$ and $w(x) = e^{-x^2}$, we get a formula to approximate

$$\int_{-\infty}^{\infty} f(x) e^{-x^2}\, dx.$$

This is *Gauss–Hermite* quadrature.

24. There are many other Gauss formulas suitable for special purposes. Most mathematical handbooks have tables of the abscissas and coefficients. The automatic generation of Gauss formulas is an interesting subject in its own right.

Numerical Differentiation

Lecture 24

Numerical Differentiation

Numerical Differentiation and Integration
Formulas from Power Series
Limitations

Numerical differentiation and integration

1. We have already noted that formulas are easier to differentiate than to integrate. When it comes to numerics the opposite is true. The graphs in Figure 24.1 suggest why.

The function of interest is represented by the straight line. The vertical bars represent the values of the function with a little error thrown in. The dashed line is the resulting piecewise linear approximation, and the area under it is the approximate integral generated by the composite trapezoidal rule.

It is clear from the figure that the errors in the individual points tend to wash out in the integral as the dashed line oscillates above and below the solid line. On the other hand, if we approximate the derivative by the slope of the dashed line between consecutive points, the results vary widely. In two instances the slope is negative!

At the end of this lecture we will see why numerical differentiation is by its nature a tricky process, but first let us derive some differentiation formulas.

Figure 24.1. *A straight line.*

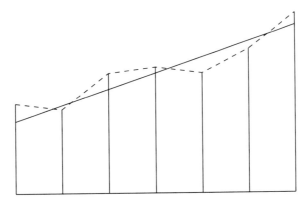

Formulas from power series

2. From elementary calculus we know that

$$f'(x) = \frac{f(x+h) - f(x)}{h} + O(h^2). \tag{24.1}$$

This suggests that we attempt to approximate the derivatives of f as a linear combination of values of f near the point x. For definiteness, we will work with the points $f(x-h)$, $f(x)$, and $f(x+h)$, where h is presumed small.

3. We begin by writing the Taylor expansion of $f(x+h)$ and $f(x-h)$ about x.

$$f(x+h) = f(x) + hf'(x) + \frac{h^2}{2}f''(x) + \frac{h^3}{6}f'''(x) + \frac{h^4}{24}f^{(4)}(x) + \cdots$$
$$f(x) = f(x)$$
$$f(x-h) = f(x) - hf'(x) + \frac{h^2}{2}f''(x) - \frac{h^3}{6}f'''(x) + \frac{h^4}{24}f^{(4)}(x) + \cdots$$

To derive a formula like (24.1), we take a linear combination of these three values that annihilates $f(x)$ on the right, leaving $f'(x)$ as its leading term. In the display below, the coefficients of the combination appear before the colons, and the result appears below the line.

$$
\begin{aligned}
1 &: f(x+h) & &= f(x) + hf'(x) + \frac{h^2}{2}f''(\xi) \\
-1 &: f(x) & &= f(x) \\
0 &: f(x-h) & &= f(x) - hf'(x) + \frac{h^2}{2}f''(\eta) \\
\hline
& f(x+h) - f(x) = & & hf'(x) + \frac{h^2}{2}f''(\xi)
\end{aligned}
$$

Note that we have replaced terms in h^2 by corresponding remainder terms. Dividing by h, we obtain the formula

$$f'(x) = \frac{f(x+h) - f(x)}{h} - \frac{h}{2}f''(\xi), \qquad \xi \in [x, x+h].$$

This formula is called a *forward-difference* approximation to the derivative because it looks forward along the x-axis to get an approximation to $f'(x)$.

4. A linear combination of $f(x-h)$, $f(x)$, and $f(x+h)$ has three degrees of freedom. In deriving the forward-difference formula, we used only two of them. By using all three, we not only can eliminate $f(x)$, but we can eliminate the terms in h^2.

Specifically,

24. Numerical Differentiation

$$
\begin{array}{lll}
1: f(x+h) & = f(x) + hf'(x) + \dfrac{h^2}{2}f''(x) + \dfrac{h^3}{6}f'''(\xi_+) \\
0: f(x) & = f(x) \\
-1: f(x-h) & = f(x) - hf'(x) + \dfrac{h^2}{2}f''(x) - \dfrac{h^3}{6}f'''(\xi_-) \\
\hline
f(x+h) - f(x-h) = & 2hf'(x) & + \dfrac{h^3}{3}\dfrac{f'''(\xi_+)+f'''(\xi_-)}{2}
\end{array}
$$

Note that the error term consists of two evaluations of f''', one at $\xi_+ \in [x, x+h]$ from truncating the series for $f(x+h)$ and the other at $\xi_- \in [x-h, x]$ from truncating the series for $f(x-h)$. If f''' is continuous, the average of these two values can be written as $f'''(\xi)$, where $\xi \in [x-h, x+h]$. Hence we have the *central-difference* formula

$$f'(x) = \frac{f(x+h) - f(x-h)}{2h} - \frac{h^2}{6}f'''(\xi), \qquad \xi \in [x-h, x+h].$$

5. Since the error in the central-difference formula is of order h^2, it is ultimately more accurate than a forward-difference scheme. And on the face of it, both require two function evaluations, so that it is no less economical. However, in many applications we will be given $f(x)$ along with x. When this happens, a forward-difference formula requires only one additional function evaluation, compared with two for a central-difference formula.

6. To get a formula for the second derivative, we choose the coefficients to pick off the first two terms of the Taylor expansion:

$$
\begin{array}{lll}
1: f(x+h) & = f(x) + hf'(x) + \dfrac{h^2}{2}f''(x) + \dfrac{h^3}{6}f'''(x) + \dfrac{h^4}{24}f^{(4)}(\xi_+) \\
-2: f(x) & = f(x) \\
1: f(x+h) & = f(x) - hf'(x) + \dfrac{h^2}{2}f''(x) - \dfrac{h^3}{6}f'''(x) + \dfrac{h^4}{24}f^{(4)}(\xi_-) \\
\hline
f(x+h) - 2f(x) + f(x-h) = & h^2 f''(x) & + \dfrac{h^4}{6}\dfrac{(f^{(4)}\xi_+)+f^{(4)}(\xi_-)}{2}
\end{array}
$$

where $\xi_+ \in [x, x+h]$ and $\xi_- \in [x, x-h]$. It follows that

$$f''(x) = \frac{f(x+h) - 2f(x) + f(x-h)}{h^2} - \frac{h^2}{6}f^{(4)}(\xi), \qquad \xi \in [x-h, x+h].$$

7. The three formulas derived above are workhorses that are put to service in many applications. However, the technique is quite flexible and can be used to derive formulas for special occasions. For instance, if we want an approximation to $f'(x)$ in terms of $f(x)$, $f(x+h)$, and $f(x+2h)$, we form the following combination:

$$
\begin{aligned}
-3 &: f(x) & &= f(x) \\
4 &: f(x+h) & &= f(x) + hf'(x) + \frac{h^2}{2}f''(x) + \frac{h^3}{6}f'''(\xi_1) \\
-1 &: f(x+h) & &= f(x) + 2hf'(x) + 2h^2 f''(x) + \frac{4h^3}{6}f'''(\xi_2)
\end{aligned}
$$

$$-3f(x) + 4f(x+h) - f(x+2h) = 2hf''(x) + \frac{2h^3}{3}f'''(\xi_1) - \frac{4h^3}{3}f'''(\xi_2)$$

Hence

$$f'(x) = \frac{-3f(x) + 4f(x+h) - f(x+2h)}{2h} + \frac{h^2}{3}[2f'''(\xi_2) - f'''(\xi_1)].$$

The error term does not depend on a single value of f'''; however, if h is small, it is approximately

$$\frac{h^2}{3} f'''(x).$$

8. The technique just described has much in common with the method of undetermined coefficients for finding integration rules. There are other, more systematic ways of deriving formulas to approximate derivatives. But this one is easy to remember if you are stranded on a desert island without a textbook.

Limitations

9. As we indicated at the beginning of this lecture, errors in the values of f can cause inaccuracies in the computed derivatives. In fact the errors do more: they place a limit on the accuracy to which a given formula can compute derivatives.

10. To see how this comes about, consider the forward-difference formula

$$D(f) = \frac{f(x+h) - f(x)}{h} + \frac{h^2}{2}f''(\xi),$$

where $D(f)$ denotes the operation of differentiating f at x. If we define the operator D_h by

$$D_h(f) = \frac{f(x+h) - f(x)}{h},$$

then the error in the forward-difference approximation is

$$D_h(f) - D(f) = -\frac{h}{2}f''(\xi).$$

In particular, if

$$|f''(t)| \leq M$$

24. Numerical Differentiation

for all t in the region of interest, then

$$|D_h(f) - D(f)| \leq \frac{M}{2}h. \tag{24.2}$$

Thus the error goes to zero with h, and we can make the approximation D_f as accurate as we like by taking h small enough.

11. In practice, however, we do not evaluate $f(t)$, but

$$\tilde{f}(t) = f(t) + e(t),$$

where $e(t)$ represents the error in computing $f(t)$. Since D_h is a linear operator, it follows that

$$D_h(\tilde{f}) - D_h(f) = D_h(e).$$

In particular, if we know that

$$|e(t)| \leq \epsilon$$

in the region of interest, then

$$|D_h(\tilde{f}) - D_h(f)| = \frac{|e(t+h) - e(t)|}{h} \leq \frac{2\epsilon}{h}.$$

Combining this inequality with (24.2), we find that the error in what we actually compute is

$$|D_h(\tilde{f}) - D(f)| \leq \frac{2\epsilon}{h} + \frac{M}{2}h.$$

12. To the extent that this bound is realistic, it says that there is a limit on how accurately we can approximate derivatives by the forward-difference formula. For large h the error formula term $Mh/2$ dominates, while for small h the term $2\epsilon/h$ dominates. The minimum occurs approximately when they are equal, that is, when $Mh/2 = 2\epsilon/h$, or

$$h = 2\sqrt{\frac{\epsilon}{M}}.$$

At this point the error bound is

$$2\sqrt{\epsilon M}.$$

For example, if ϵ equals the rounding unit ϵ_M and $M = 1$, then the minimum error is $2\sqrt{\epsilon_M}$. In other words, we cannot expect more than half machine precision in such a forward difference.

13. To illustrate this point, the MATLAB code

```
for i=1:14,
    x(i) = (sin(pi/3.2+10^(-i))-sin(pi/3.2))/10^(-i);
    y(i) = x(i) - cos(pi/3.2);
end
```

uses forward differencing to compute the derivative of $\sin(\pi/3.2)$ with $h = 10^{-1}, \ldots, 10^{-14}$. The array y contains the error in these approximations. Here is the output.

x	y
0.51310589790214	−0.04246433511746
0.55140366014496	−0.00416657287465
0.55515440565301	−0.00041582736659
0.55552865861230	−0.00004157440730
0.55556607565510	−0.00000415736450
0.55556981726212	−0.00000041575748
0.55557019096319	−0.00000004205641
0.55557023426189	0.00000000124229
0.55557025646635	0.00000002344675
0.55557003442175	−0.00000019859785
0.55556670375267	−0.00000352926693
0.55555560152243	−0.00001463149717

The error decreases until $h = 10^{-8}$ and then increases. Since ϵ_M is 10^{-14}, this is in rough agreement with our analysis.

14. It is worth noting that if you turn the output sideways, the nonzero digits of y plot a graph of the logarithm of the error. The slope is the same going down and coming up, again as predicted by our analysis.

15. It is important not to get an exaggerated fear of numerical differentiation. It *is* an inherently sensitive procedure. But as the above example shows, we can often get a good many digits of accuracy before it breaks down, and this accuracy is often sufficient for the purposes at hand.

Bibliogaphy

Introduction
References

Introduction

1. The following references fall into three classes. The first consists of elementary books on numerical analysis and programming. The number of such books is legion, and I have listed far less than a tithe. The second class consists of books treating the individual topics in these notes. They are a good source of additional references. I have generally avoided advanced treatises (hence the absence of Wilkinson's magisterial *Algebraic Eigenvalue Problem*). The third class consists of books on packages of programs, principally from numerical linear algebra. They illustrate how things are done by people who know how to do them.

2. No bibliography in numerical analysis would be complete without referencing Netlib, an extensive collection of numerical programs available through the Internet. Its URL is

>   http://www.netlib.org

The Guide to Available Mathematical Software (GAMS) at

>   http://gams.nist.gov

contains pointers to additional programs and packages. Be warned that the preferred way to access the Internet changes frequently, and by the time you read this you may have to find another way to access Netlib or GAMS.

References

E. Anderson, Z. Bai, C. Bischof, J. Demmel, J. Dongarra, J. Du Croz, A. Greenbaum, S. Hammarling, A. McKenney, S. Ostrouchov, and D. Sorensen. *LAPACK Users' Guide*. Society for Industrial and Applied Mathematics, Philadelphia, PA, second edition, 1995.

K. E. Atkinson. *An Introduction to Numerical Analysis*. John Wiley, New York, 1978.

Å. Björck. *Numerical Methods for Least Squares Problems*. Society for Industrial and Applied Mathematics, Philadelphia, PA, 1994.

R. P. Brent. *Algorithms for Minimization without Derivatives*. Prentice–Hall, Englewood Cliffs, NJ, 1973.

T. Coleman and C. Van Loan. *Handbook for Matrix Computations.* Society for Industrial and Applied Mathematics, Philadelphia, PA 1988.

S. D. Conte and C. de Boor. *Elementary Numerical Analysis: An Algorithmic Approach.* McGraw–Hill, New York, third edition, 1980.

G. Dahlquist and Å. Björck. *Numerical Methods.* Prentice–Hall, Englewood Cliffs, NJ, 1974.

B. N. Datta. *Numerical Linear Algebra and Applications.* Brooks/Cole, Pacific Grove, CA, 1995.

P. J. Davis. *Interpolation and Approximation.* Blaisdell, New York, 1961. Reprinted by Dover, New York, 1975.

P. J. Davis and P. Rabinowitz. *Methods of Numerical Integration.* Academic Press, New York, 1967.

J. E. Dennis and R. B. Schnabel. *Numerical Methods for Unconstrained Optimization and Nonlinear Equations.* Prentice–Hall, Englewood Cliffs, NJ, 1983.

J. J. Dongarra, J. R. Bunch, C. B. Moler, and G. W. Stewart. *LINPACK User's Guide.* Society for Industrial and Applied Mathematics, Philadelphia, PA, 1979.

L. Eldén and L. Wittmeyer-Koch. *Numerical Analysis: An Introduction.* Academic Press, New York, 1993.

G. Evans. *Practical Numerical Integration.* Wiley, New York, 1990.

G. E. Forsythe, M. A. Malcolm, and C. B. Moler. *Computer Methods for Mathematical Computations.* Prentice–Hall, Englewood Cliffs, NJ, 1977.

C.-E. Fröberg. *Introduction to Numerical Analysis.* Adison–Wesley, Reading, MA, 1969.

G. H. Golub and C. F. Van Loan. *Matrix Computations.* Johns Hopkins University Press, Baltimore, MD, second edition, 1989.

N. J. Higham. *Accuracy and Stability of Numerical Algorithms.* Society for Industrial and Applied Mathematics, Philadelphia, PA, 1995.

A. S. Householder. *The Numerical Treatment of a Single Nonlinear Equation.* McGraw–Hill, New York, 1970.

E. Isaacson and H. B. Keller. *Analysis of Numerical Methods.* John Wiley, New York, 1966.

D. Kahaner, S. Nash, and C. Moler. *Numerical Methods and Software.* Prentice–Hall, Englewood Cliffs, NJ, 1988.

B. W. Kernighan and D. M. Ritchie. *The C Programming Language.* Prentice–Hall, Englewood Cliffs, NJ, 1988.

C. L. Lawson and R. J. Hanson. *Solving Least Squares Problems.* Society for Industrial and Applied Mathematics, Philadelphia, PA, 1995.

M. J. Maron and R. J. Lopez. *Numerical Analysis: A Practical Approach.* Wadsworth, Belmont, CA, third edition, 1991.

C. Moler, J. Little, and S. Bangert. *Pro-Matlab User's Guide.* The Math Works, Sherborn, MA, 1987.

J. M. Ortega. *Numerical Analysis: A Second Course.* Academic Press, New York, 1972.

P. H. Sterbenz. *Floating-Point Computation.* Prentice–Hall, Englewood Cliffs, NJ, 1974.

G. W. Stewart. *Introduction to Matrix Computations.* Academic Press, New York, 1973.

J. Stoer and R. Bulirsch. *Introduction to Numerical Analysis.* Springer-Verlag, New York, second edition, 1993.

D. S. Watkins. *Fundamentals of Matrix Computations.* John Wiley & Sons, New York, 1991.

J. H. Wilkinson. *Rounding Errors in Algebraic Processes.* Prentice–Hall, Englewood Cliffs, NJ, 1963.

J. H. Wilkinson and C. Reinsch. *Handbook for Automatic Computation. Vol.* II: *Linear Algebra.* Springer-Verlag, New York, 1971.

Index

Italics signify a defining entry. The abbreviation *qv* (*quod vide*) means to look for the topic in question as a main entry. The letter "n" indicates a footnote.

absolute error, 7
    as convergence criterion, 7
absolute value
    as a norm, 113
astrologer, 150

back substitution, 100
backward error analysis, *55–57*, 125, 128
    backward stability, 55
    Gaussian elimination, 122–124
    stable algorithm, 55
    sum of n numbers, 53–55
    sum of two numbers, 50–51
backward stablity, *see* backward error analysis
balanced matrix, 120, 136
base, *see* floating-point arithmetic
basic linear algebra subprogram, *see* BLAS
binary arithmetic, *see* floating-point arithmetic
BLAS, *86–88*
    `axpy`, 87, 98n, 109
    `copy`, 88
    `dot`, 87
    `ger`, 109
    `imax`, 108
    in Gaussian elimination, 108–109
    level-three, 109n
    level-two, *109*
    `scal`, 88, 109
    `swap`, 108
bracket for a root, *see* nonlinear equations

C, 69, 80n, 86
    storage of arrays, 83, 109
cache memory, 86
cancellation, 61–63, 65, 176
    death before cancellation, 61–62, 104
    in a sum of three numbers, 61
    in ill-conditioned problems, 63
    in the quadratic formula, 62
    recovery from, 62–63, 65
    revealing loss of information, 61, 62
characteristic, *see* floating-point arithmetic
Chebyshev points, *152*
Cholesky algorithm, 90–95, 98
    and rounding error, 98
    and solution of linear systems, 95
    Cholesky decomposition, *90*
    Cholesky factor, *90*
    Cholesky factorization, *90*
    column orientation, 94–95
    economization of operations, 92
    economization of storage, 92
    equivalence of inner and outer product forms, 98
    implementation, 92–95
    inner product form, 97–98
    operation count, 95, 98
    relation to Gaussian elimination, 91–92
column index, *see* matrix
column orientation, 84–85, 87, 94, 98n

and level-two BLAS, 109
  general observations, 86
column vector, *see* vector
component of a vector, *see* vector
computer arithmetic, *see*
    floating-point arithmetic
condition
  artificial ill-conditioning,
    120–121
  condition number, *41*, 42
  condition number with respect
    to inversion, 120
  generating matrices of known
    condition, 129
  ill-conditioning, 41, 51, 57, 63,
    65
  ill-conditioning of
    Vandermonde, 136
  ill-conditioning revealed by
    cancellation, 63
  linear system, 119–120, 169
  roots of nonlinear equations,
    41–42
  roots of quadratic equations,
    63
  sum of n numbers, 57
  well-conditioning, 41
condition number, *see* condition
conformity, *see* partitioning,
    product, sum
constant slope method, *17*, 22
  as successive-substitution
    method, 23
  convergence analysis, 18–19
  failure, 19
  linear convergence, 19
convergence, *see* convergence of
    order p, cubic
    convergence, etc.
convergence of order p, *20*, 22
  and significant figures, 20
  limitations for large p, 21

multipoint methods, 33
nonintegever orders, 21
two-point methods, 32
Cramer's rule, 127, *130*
  compared with Gaussian
    elimination, 130–131
cubic convergence, 20, 21

decimal arithmetic, *see*
    floating-point arithmetic
determinant, 77
  unsuitability as measure of
    singularity, 77–78
diagonal matrix, *71*
  two-norm of, 127
difference equation, 32
  effects of rounding error, 63–65
difference quotient, 27
differentiation
  compared with integration,
    157, 181
dimension of a vector, *see* vector
divided difference, 144–146
  and error in interpolant, 148
  as coefficient of Newton
    interpolant, 144
  computation, 145–146
  operation count, 146
  relation to derivative, 149

economist, *see* astrologer
element of a matrix, *see* matrix
elementary permutation, *105*
error bounds for floating-point
    operations, *see* rounding
    error
error function, 157
Euclidean length of a vector, *see*
    norm, two-norm
exponent, *see* floating-point
    arithmetic

fixed point, *21*

Index

fixed-point number, 45
floating-point arithmetic
    advantage over fixed-point arithmetic, 45
    avoidance of overflow, 47–48
    base, 45–46
    binary, 46, 49
    characteristic, 45n
    decimal, 46
    double precision, *46*
    error bounds for floating-point operations, *see* rounding error
    exponent exception, 47
    floating-point number, *45–46*
    floating-point operations, 49–50
    fraction, 45
    guard digit, 50
    hexadecimal, 46
    high relative error in subtraction, 50
    IEEE standard, *46*, 47, 50
    mantissa, 45n
    normalization, *45–46*
    overflow, *47–48*, 141, 144
    quadruple precision, *46*
    rounding error, *qv*
    single precision, *46*
    underflow, *47–48*, 141, 144
    unnormalized number, 45
FORTRAN, 94
    storage of arrays, 84, 109
forward-substitution algorithm, 80
fraction, *see* floating-point arithmetic

GAMS, 187
Gauss, C. F., 31, 169, 176n
Gaussian elimination, *92*, *98–100*, 108, 127, 128, 135
    and LU decomposition, 100–102
    and the Cholesky algorithm, 91–92
    back substitution, 100
    backward error analysis, 122–124
    BLAS in implementation, 108–109
    compared with Cramer's rule, 130–131
    compared with invert-and-multiply, 130
    complete pivoting, 105n
    exponential growth, 125–126
    for Hessenberg matrices, 110–111
    for tridiagonal matrices, 111
    general assessment, 126
    growth factor, *125*
    implementation, 102
    multiplier, *99*, 101–102, 122, 123
    operation count, 102, 111
    partial pivoting, 104–108, 125
    pivot, *103*
    pivoting, 103–108, 111
    stability, 125
Gaussian quadrature, 167
    convergence, 176
    derivation, 175
    error formula, 176
    Gauss–Hermite quadrature, 177
    Gauss–Laguerre quadrature, 176
    Gauss–Legendre quadrature, 176
    introduction, 169
    positivity of coefficients, 175–176
guard digit, *see* floating-point arithmetic

Hessenberg matrix, *110*

and Gaussian elimination,
110–111
hexadecimal arithmetic, *see*
floating-point arithmetic

identity matrix, *see* matrix
IEEE standard, *see* floating-point
arithmetic
ill-conditioning, 123
inner product, 98n, 171
Cholesky algorithm, 97–98
computed by BLAS, 87
integral
definite, 157
indefinite, 157
integration
compared with differentiation,
157, 181
intermediate value theorem, 4
interpolation, 35
interpolatory method, 34
interval bisection, *4–6*
combined with secant method,
37–40
convergence, 6
implementation, 5–6
inverse matrix, *78*, 105, 120n, 130
and linear systems, 78
calculation of, 95
lower triangular matrix, 101
of a product, 78
of a transpose, 78
invert-and-multiply algorithm, *see*
linear system

Lagrange interpolation, 144
Lagrange polynomials, *137–138*,
161
least squares, 89, 176
Legendre, A. M., 176n
linear combination, 75
linear convergence, 19, *20*, 22
rate of convergence, *20*

linear equations, *see* linear system
linear fractional method, 34–36
linear independence, 77
linear system, 77, 104
and partial pivoting, 105–108
and relative residual, 128–129
and rounding error, 120
artificial ill-conditioning,
120–121
condition, 119–120
Cramer's rule, *qv*
existence of solutions, 77
invert-and-multiply algorithm,
78, 95, 127, 130
lower triangular system,
79–81, 84–85, 90, 97
matrix representation, 72
nonsymmetric system, 98
operation count, *qv*
perturbation analysis, 116–117
positive-definite system, 89, 95
solution by LU decomposition,
78–79, 89
triangular system, 95
uniqueness of solutions, 77
locality of reference, *83*, 85, 86
lower triangular matrix, *79*, 79
inverse, 101
product, 101
unit lower triangular matrix,
102
lower triangular system
nonsingularity, 143
LU decomposition, *79*, 98, 102,
120n, 123
and Gaussian elimination,
100–102
existence, 103
solution of linear systems,
78–79, 89
with partial pivoting, 106–107

mantissa, *see* floating-point

Index   195

arithmetic
MATLAB, 127
matrix, *69*
    column index, *69*
    diagonal matrix, *qv*
    element, *69*
    identity matrix, *71*
    inverse matrix, *qv*
    lower triangular matrix, *qv*
    nonsingular matrix, *qv*
    normally distributed matrix, 129
    operations, *see* multiplication by a scalar, sum, product, transpose
    order, *69*
    orthogonal matrix, *qv*
    partitioning, *qv*
    positive definite matrix, *qv*
    rank-one matrix, *qv*
    represented by upper-case letters, 70
    row index, *69*
    symmetric matrix, *89*
    unit lower triangular matrix, 102
    upper triangular matrix, 79, *79*
    Vandermonde matrix, *qv*
    zero matrix, *70*
matrix perturbation theory, 116
monic polynomial, *172*
Muller's method, 33–34, 135
multiple zero, *24*
    behavior of Newton's method, 24
multiplication by a scalar
    and norms, 113
    and transposition, 72
    matrix, *70*
    vector, 71
multipoint method, *33*, 34, 135

rate of convergence, 33
natural basis interpolation, 135–137, 141
    evaluation by synthetic division, 141–142
    ill-conditioning of Vandermonde, 136
    Vandermonde matrix, *qv*
Netlib, 187
Newton interpolation, *142*, 144
    addition of new points, 144
    coefficients as divided differences, 144
    computation of coefficients, 145–146
    evaluation by synthetic division, 142
    existence and uniqueness, 143–144
Newton's method, *9–15*, 22, 34
    analytic derivation, 10
    as confluent case of secant method, 29
    as successive-substitution method, 23
    calculating reciprocals, 11
    calculating square roots, 11
    convergence analysis, 12–14, 19
    convergence to multiple zero, 24
    derivative evaluation, 4, 17
    divergence, 11
    failure, 37
    geometric derivation, 9
    quadratic convergence, 14, 19, 23
    retarded convergence, 14–15, 25
    starting values, 15
nonlinear equations, 4
    analytic solutions, 4

bracket for a root, 5
condition of roots, see
    condition, roots of
    nonlinear equations
effects of rounding error, 6
errors in the function, 37,
    40–42
existence of solutions, 4, 5
general observations, 3–4
hybrid of secant method and
    interval bisection, *37–40*
interpolatory method, 33
interval bisection, *qv*
Muller's method, *qv*
multipoint method, *qv*
Newton's method, *qv*
polynomial equations, 25
quadratic formula, *qv*
quasi-Newton method, 4, 17n,
    17, 27
root and zero contrasted, 9
secant method, *qv*
successive substitutions
    method, *qv*
two-point method, *qv*
uniqueness of solutions, 4
nonsingular matrix, *78*, 90
    perturbation of, 117
norm
    and orthogonal matrices, 127
    column-sum norm, 115
    consistency, *114*, 116, 127
    Euclidean norm, 114
    Frobenius norm, *115*
    infinity-norm, *113–115*
    Manhattan norm, 114
    matrix norm, *114–115*
    max norm, 114
    normwise relative error,
        115–116, 117, 120, 128
    of a diagonal matrix, 127
    of a rank-one matrix, 127

one-norm, *113–115*
row-sum norm, 115
triangle inequality, 114
triangle inequality., 113
two-norm, 72, *113–114*, 115n,
    127–128
vector norm, *113–114*
numerical differentiation
    central-difference formula, 183
    compared with numercial
        integration, 181
    error analysis, 184–186
    forward-difference formula,
        182, 184
    second derivative, 183
    three-point
        backward-difference
        formula, 184
numerical integration, 41, *157*
    and Lagrange polynomials,
        161–162
    change of intervals, 158
    compared with numerical
        differentiation, 181
    Gauss–Hermite quadrature,
        177
    Gauss–Laguerre quadrature,
        176
    Gauss–Legendre quadrature,
        176
    Gaussian quadrature, *qv*
    Newton–Cotes formulas,
        161–162, 166, 167
    Simpson's rule, *qv*
    trapezoidal rule, *qv*
    treatment of singularities,
        167–168
    undetermined coefficients,
        162–163, 167, 169
    weight function, *167*, 176
numerical quadrature, 157

operation count, 81–82

approximation by integrals,
    81, 95
Cholesky algorithm, 95
divided difference, 146
Gaussian elimination, 102
Gaussian elimination for
    Hessenberg matrices, 111
Gaussian elimination for
    tridiagonal matrices, 111
interpretation and caveats,
    81–82
lower triangular system, 81
synthetic division, 142
order of a matrix, *see* matrix
orthogonal function, *171*
orthogonal matrix, *127*
and two-norm, 127
random, 129
orthogonal polynomials, 169, *171*
existence, 173–174
Legendre polynomials, 176
normalization, 172
orthogonality to polynomials
    of lesser degree, 172
reality of roots, 174–175
three-term recurrence, 174
outer product, 97
overflow, *see* floating-point
    arithmetic
overwriting, 92, 93, 102

partitioning, *74*, 90
by columns, 74
conformity, 74
matrix operations, 74
paritioned sum, 74
partitioned product, 74
positive-definite matrix, 90
perturbation analysis, *57*
linear system, 116–117
sum of n numbers, 57
pivoting, *see* Gaussian elimination
polynomial

bases, 143n, 172
evaluation by synthetic
    division, 141–142
monic, *172*
number of distinct zeros, 138
polynomial interpolation, *137*
approximation to the sine, 150
at Chebyshev points, 151–153
convergence, 150–153, 161
error bounds, 149–150
error in interpolant, 147–149
existence, 137–138
extrapolation and
    interpolation, 149–150
failure of convergence, 151, 153
general features, 137
Lagrange interpolation,
    137–138, 141
linear interpolation, 149
natural basis interpolation, *qv*
Newton interpolation, *qv*
quadratic interpolation,
    135–136
Runge's example, 151
shift of origin, 136
uniqueness, 138–139
Vandermonde matrix, *qv*
positive-definite matrix, *89*
calculation of inverse, 95
nonsingularity, 90
partitioned, 90
without symmetry
    requirement, 89n
precision, *see* floating-point
    arithmetic
product, 70
and transposition, 72
associativity, 72
conformity, 71
distributivity, 72
inner product, 72
inverse of, 78

matrix, *70–71*
    matrix-vector, *71*, 75
    noncommutativity of matrix product, 72, 74
    of partitioned matrices, 74
    of triangular matrices, 101
    rank-one matrix and a vector, 73
    recipe for matrix product, 71

quadratic convergence, *14*, 19–20
    doubling of significant figures, 14
    of Newton's method, 14, 19
quadratic formula, 61–63
    discriminant, 63
    revised, 63

rank-one matrix, *73*
    computing with, 73
    storing, 73
    two-norm of, 127
reciprocal calculated by Newton's method, 11
regression, 89
relative error, *7*, 57, 128
    and significant figures, 7–8
    as convergence criterion, 8
    normwise, *see* norm
    and rounding error, 49
relative residual, *128*
    and stability of linear systems, 128–129
residual, *128*
rounding error, 40, 47, *48–49*
    accumulation, 55, 58–59, 65
    adjusted rounding unit, 55
    cancellation, *qv*
    chopping, 48–49
    computation of the rounding unit, 49
    difference equation, 63–65
    error bounds, *48–49*, 59
    error bounds for floating-point operations, *50*
    general observations, 65–66
    in linear systems, 120
    inferiority of chopped arithmetic, 58–59
    machine epsilon, 49
    magnification, 65
    and relative error, 49
    rounding, 48
    rounding unit, *49*, 120, 123
    statistical analysis, 59
    truncation, 48
rounding unit, *see* rounding error
rounding-error analysis
    accuracy of a sum of positive numbers, 58–59
    accuracy of computed sum, 58
    backward error analysis, *qv*
    cancellation, *qv*
    Cholesky algorithm, 98
    difference equation, 64–65
    Gaussian elimination, *qv*
    numerical differentiation, 184–186
    simplification of error bounds, 54–55
    single error strategy, 65
    sum of n numbers, 53–55
    sum of two numbers, 50–51
row index, *see* matrix
row orientation, 83–84, 87, 98n
    and level-two BLAS, 109
    general observations, 86
row vector, *see* vector

scalar, *70*
    as 1×1 matrix, 69
    multiplication by, 70
    represented by lower-case Latin or Greek letter, 70
secant method, *27*, 34
    as an interpolatory method, 33

Index

combined with interval
bisection, *37–40*
convergence, see two-point
method, 21
failure, 34, 37
geometric derivation, 28
Newton's method as confluent
case, 29
quasi-Newton method, 17
significant figures
and quadratic convergence, 14
and relative error, 7–8
simple zero, 24
Simpson's rule, *158*
as a partial Gauss quadrature,
169
composite rule, *165–166*
derived by undetermined
coefficients, 162–163
error formula, 166
error in composite rule, 167
exact for cubics, 167, 169
half-simp rule, 166
singular matrix, 123
spline interpolant, 153
square root
calculated by Newton's
method, 11
stable algorithm, 41, 55, 128–129
backward error analysis, qv
dialogue on stability, 55–56
Gaussian elimination, 125
synthetic division, 142
Stewart, G. W., 103
sublinear convergence, 20n
successive substitution method, *21*
convergence, 21
geometric interpretation, 21
sum, 70
and transposition, 72
associativity, 72
comformity, 70

commutativity, 72
distributivity, 72
matrix, *70*
of partitioned matrices, 74
vector, 71
superlinear convergence, *20*
not of order p, 21n
two-point methods, 32
symmetric matrix
economization of operations,
92
synthetic division, *141–142*
evaluation of Newton
interpolant, 142
operation count, 142
stability, 142

transpose, 69–70, *72*
and matrix operations, 72
inverse of, 78
trapezoidal rule, *158–160*
analytic derivation, 159
composite rule, *160*, 181
error formula, 159–160
error in the composite rule,
160–161
geometric derivation, 158–159
triangle inequality, see norm
tridiagonal matrix, 111
and Gaussian elimination, 111
two-point method, *28*
convergence analysis, 29–32
rate of convergence, 32

underflow, see floating-point
arithmetic

Vandermonde matrix, *137*, 144
ill-conditioning, 136
nonsingularity, 135, 138
vector, *69*
column vector, 69
component, *69*

dimension, 69
as $n \times 1$ matrix, 69
n-vector, 69
represented by lower-case
    Latin letters, 70
row vector, *69*, 72
vector operations, *see*
    multiplication by a scalar,
    sum, product, transpose
vector supercomputer, 109
virtual memory, *83*, 84, 86
    page, *83*
    page hit, *83*
    page miss, *83*, 83–85

Weierstrass approximation
    theorem, 176
weight function, 171
Wilkinson, J. H., 37, 110

zero matrix, *see* matrix
zero of a function, 9